# ENCOUNTERS WITH
# MUSLIM NOMADS

The Wandering People who have
Shaped the Islamic World

Written and Illustrated by
## Luqman Nagy

© Text and illustrations Luqman Nagy 1445 AH/ 2024 CE

First Published in March 2024 by:

**Ta-Ha Publishers Ltd,**
Unit 4, The Windsor Centre,
Windsor Grove, West Norwood,
London, SE27 9NT
United Kingdom

www.**tahapublishers.com**
support@tahapublishers.com

All rights reserved. No part of this publication may be reproduced, stored in any retrieval system, or transmitted in any form or by any means, electronic, mechanical, photocopying, recording or otherwise, without the prior written permission of the publishers.

Written and Illustrated by: **Luqman Nagy**
Editor: **Arooj Rashid Hussain**
Cover & Book Design by: **Shakir Abdulcadir**

A catalogue record of this book is available from the British Library
ISBN: 978-1-915357-19-9 (Paperback)


Printed and bound by: Mega Basim, Türkiye
10 9 8 7 6 5 4 3 2 1

This book is dedicated to practical Muslim Nomads and their age-old traditions that brilliantly shine on these pages.

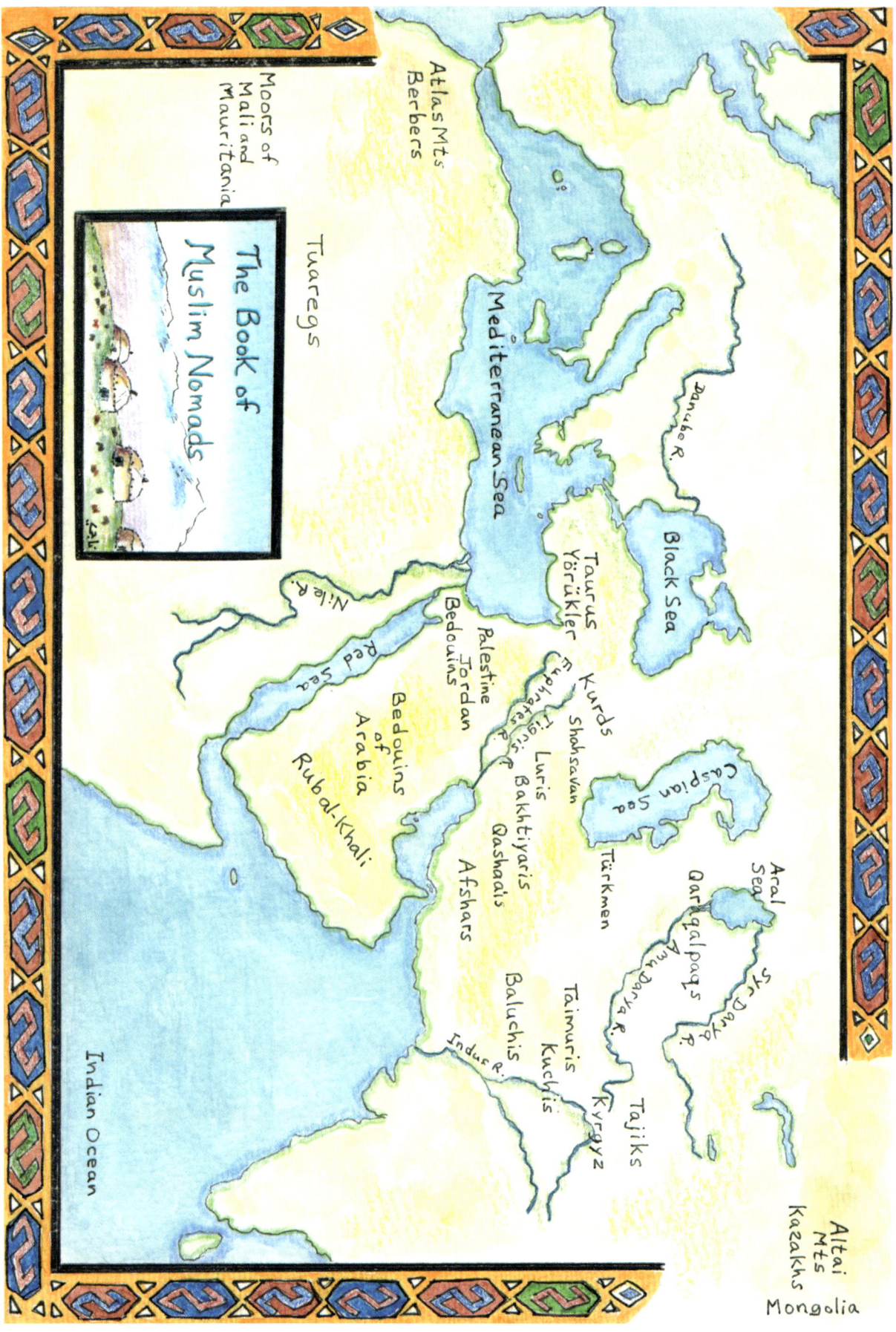

# CONTENTS

|  | Author's Note | 7 |
|---|---|---|
|  | Preface | 13 |
| 1 | Who is a Nomad? | 19 |
| 2 | Ibn Khaldun and Bedouin Civilisation | 23 |
| 3 | Berbers of the Atlas Mountains | 29 |
| 4 | Tuaregs of the Sahara Desert | 39 |
| 5 | Yörükler of the Taurus Mountains | 51 |
| 6 | Bedouins of Arabia | 61 |
| 7 | Bakhtiyaris of Western Iran | 77 |
| 8 | Qashqa'is of Southern Iran | 91 |
| 9 | Kuchis of Southern Afghanistan | 103 |
| 10 | Kyrgyz of Northeastern Afghanistan | 115 |
| 11 | Altai Kazakhs of Northwestern Mongolia | 133 |
| 12 | Future of Nomadism | 147 |
|  | Select Bibliography | 155 |

'Abdullah bin 'Umar ؓ narrated that Allah's Messenger ﷺ said:
"Be in this world as though you were a stranger or a traveller."

(Bukhari)

'Nomads have, as long as we know, represented a threat and a source of wonder to settled populations… The tension between mobility and stasis, in terms of freedom and security, and the fantasies of an independent, free-floating existence, have perhaps always been part of European settled populations' understanding of themselves and the 'other' (Peters, 2006). The figure of the nomad as the embodiment of freedom and irresponsibility and a challenge to the order of things is thus deeply embedded in European understandings of mobility and stasis. The threatening image of mobile peoples as destroyers of order and progress is as old as the romantic fantasies.'

(Ada Ingrid Engebrigtsen, 2017)

# AUTHOR'S NOTE

Travel is proverbially all about the people one
meets; the nomads who crossed my path during
a half century of transversing North Africa,
the Middle East and Central Asia, have granted
me some of my richest and most rewarding
experiences. The genesis of this book began by my
reflection on these important encounters.

'Those scattered beyond the boundaries of the city have long been the subject of puzzlement and fantasy by those within its walls, as well as a metaphor for ways of being in the world.'
(Peters, 2006)

One newly coined phrase of today's youthful Generation Z is the ubiquitous 'digital nomad'. As many digital nomads are now choosing to reside in Muslim lands like Morocco and Turkey, one wonders if there might be an occasional, serendipitous encounter between them and the authentic indigenous nomads…

Educated in Vancouver, B.C. on the west coast of Canada, I grew up on a diet of travelogues by authors like the American adventurer, Richard Haliburton, and the intrepid British scholar, Freya Stark. Both authors frequently mentioned their fascinating encounters with local nomadic tribes. After completing my undergraduate education in Modern Foreign Languages and History in 1970, I felt adequately prepared for an extended sojourn abroad. Western Europe proved too tame for my liking, so I headed south and eventually settled down in Lisbon, Portugal, at the time, the last remaining colonial power in Europe. It was in Lisbon, listening to the nightly BBC shortwave radio broadcasts, that I stumbled upon what I believed to be an obvious Arabic language channel, the voice of 'Radio Rabat' (*sawt ar-rabat*). The uniqueness of the language only increased my desire to visit Portugal's nearest African neighbour. During the winter break (1970-71), I ventured across the narrow Strait of Gibraltar to discover Morocco, the most westerly outpost of the classical Muslim world. At the time, the Kingdom of Morocco was still an extremely traditional country, whose mosques were only open to the faithful. As a young Canadian, this was my first encounter with Muslims and their heedful religion. Most impressive for me was how the Islamic call to prayer (*adhan*) seemed to physically regulate the lives of everyone I met. Unable to converse with locals in their vernacular, I promised myself that one day I would learn Arabic, the language I believed to be the key to unlocking the secrets of this outwardly beautiful, but still very mysterious, religion and its followers. Amongst the sedentary mountain folk of the High Atlas Mountains of Morocco were small bands of migrating Berber families. Their apparent wanderlust instantly appealed to the 'nomad' in myself.

In the 1970s, as a young budget-conscious university graduate and researcher, I often travelled by dubious public transport; in northeastern Afghanistan, I rode on horseback later venturing on foot along the well-trodden paths of the semi-nomadic Wakhi locals. In this book, I have included two photographs taken during my time in the Wakhan Corridor. After a full decade of dedicated travel and the completion of graduate degrees, I became a lecturer at King Fahd University of Petroleum and Minerals (Dhahran, Saudi Arabia). This location enabled me to easily travel throughout the entire Islamic world.

Readers of this book might wonder why I describe my encounters in this book in the third person, and use drawings as opposed to photographs for illustrations. Even the chapters of my book follow a geographical progression – from the Berber nomads of northwest Africa to the nomads of Iran and Afghanistan further to the east – rather than a personal one. My intention in researching Muslim nomads was never to write a book. This independent research was a genuine hobby and an extension of my interest in all things Islamic: Arabic calligraphy, tile making, architecture, history, numismatics, to name a few. As a young traveller, I was simply awestruck by the nomads' sustainability ethos, and instantly recognised it as something worthy of respect and admiration. After all, I was born in Vancouver, where I personally witnessed the birth of Greenpeace, the global environmental movement. Since the late 1960s, I have supported environmental awareness, and as a new Muslim in the early 1970s, I became thoroughly convinced of the importance of our blessed role as custodians of this Earth. Muslim nomads strike me as having truly understood this role. As my own research developed, and news of the concerning impact of stationary nations on the planet gained momentum, I realised that my private honouring of these tribes should be put to good use. The very act of learning – and in some cases, revising our own perceptions – about the honourable lifestyles of Muslim nomads could offer a sound antidote to our increasingly consumerist and unpredictable way of life. I hope this book offers our readers just that. I wanted to make sure this travel book shifts the focus from me to those nomads that society has for so long seen on the peripheries of civilisation.

As a young, sometimes foolhardy traveller, I did take unnecessary risks. For example, in preparing for my trek along the remote Wakhan Corridor of northeastern Afghanistan, I thoughtlessly purchased a bottle of Aspirin as my only medicine. This proved totally inadequate when I contracted typhoid fever from drinking the polluted water of the Panja River. I only recovered when a young German linguist conducting field research in Pamiri languages heard of my plight when passing through Fayzabad, the capital of Badakhshan province. He had prepared much better for his stay in the mountains by packing plenty of antibiotics! Despite my unpreparedness and limited access to a camera throughout my travels, my encounters of the nomadic encampments were vivid and enlightening. I therefore preferred to paint illustrations of my visits as a personal homage to these noble Muslims, who always seemed preserved in my peripheral vision.

As a septuagenarian, my life has been fulfilling and rewarding. I pray that future generations may live in a more tolerant world, in which the Muslim nomads might finally be recognised for their unheralded contributions to the survival of humankind. Nomads, despite the vicissitudes of their often precarious existence, remain threads in the fabric of the societies they nominally inhabit. In the twenty-first century, we are truly blessed to still have in our midst pockets of these itinerant wayfarers. The eco-friendly lifestyle, traditions, and exquisite material culture of such humble Muslims continue to engender a reverence for them. Ideally, I wish my prospective readers could appreciate these nomads for whom they really are: incredibly ingenious, skilled, God-fearing pastoralists who practise a minimalist, eco-friendly lifestyle from which much can be learnt.

Finally, I hope that the chapters of this book may increase awareness of these fragile, long-marginalised, minority populations that still inhabit the remoter regions of our ever-shrinking world. The nomads' sustainability ethos alone is laudable and is precisely what has been their attraction for me. Indeed, for countless generations they have walked gently on God's green Earth as its respectful custodians.

And the (true) servants of the Most Gracious (Allah) are those who walk gently upon the earth… [25:63]

# PREFACE

While today the term 'nomad' most likely connotes an aimless wanderer, for thousands of years tribes have roamed the deserts and steppes of the world for a very specific purpose: procuring sustenance for themselves and their livestock. This often meant hunting game, gathering seasonal food such as berries, nuts and fruit and, of course, locating lush pasture for their animals.

"The seasonal nomadic migration of people and animals as a distinctive way of life and specialised mode of production involving substantial populations most likely goes back some six to seven thousand years in the Near East ... Animals and migration were the cornerstone of the nomadic pastoralist lifestyle."[1]

Nomadism in its purist form demanded perpetual mobility in order to support vast herds of animals. The grasslands of the 'unharvested steppe', stretching from Mongolia to Central Asia and further west, are frequently cited as being the homeland of one of history's greatest nomadic peoples. Here, in the thirteenth century, Genghis Khan created the largest contiguous land empire ever seen in history. Mongol hordes poured out of the east moving west in search of pasture for their steeds – the backbone of their mighty army.[2] The cultivated farmland of the civilised world prevented the Mongols' unhindered access to the vital ocean of steppe grassland.[3] Genghis Khan reportedly despised cities and the surrounding farmland (with its ingenious irrigation channels) because they denied him grazing land for his vast army of mounted soldiers.[4]

1. Wertime, *Sumak Bags of Northwest Iran and Transcaucasia*, 14.

2. Pastoral nomads, like the Mongols, are sometimes marginalised by historians as they were considered 'barbarians', or even 'outside of history' as Arnold Toynbee, the British historian, once remarked. However, the fact is that they, along with many other nomadic peoples, became a significant driving force in history. For example, in the Islamic era, the Seljuqs (tenth to early thirteenth century CE) were crucial in the Islamisation of modern-day Türkiye. The Seljuqs were originally a group of central Asian Turkic nomads, who on becoming Muslims, eventually became guardians of the weakened Abbasid Caliphate in Baghdad, which they seized in 1055 CE. From there, they ruled an empire (the Great Seljuq Sultanate) that included parts of central Asia, Iran, Iraq and Syria. An offshoot of this dynasty, the Seljuqs of Rum, defeated the Byzantine emperor at Malazgirt in eastern Türkiye in 1071 CE. This victory enabled thousands of Seljuq nomads to flood into Anatolia (Asia Minor, or modern-day Asian Türkiye), thus making the once Byzantine Christian land forever Muslim. The Seljuqs also helped prevent the Fatimid Shiite form of Islam from predominating in the Middle East. The Ottoman Empire (1299-1923) had its genesis in the late Seljuq period. The ancestors of the Ottomans, the Oghuz Turks, were initially vassals of the waning Seljuqs of Rum and when that state finally ceased to exist, they filled the power vacuum eventually dominating Anatolia. The Ottomans are, therefore, often cited as the successors to the Seljuqs. In Iran, the rulers of the Zand dynasty (1750-1794), whose capital was Shiraz, were descended from a nomadic Luri tribe. The Zand rulers were ousted, in turn, by the Qajars (1794-1925), a Turkic tribe of pastoral nomads from northern Iran. In 'the far west' of the Islamic world, two noteworthy Islamic dynasties were founded by Berber nomads: the Almoravids (1050-1147) and the Almohads (1147-1248). Both dynasties had empires, which included Islamic Spain, present-day Morocco and parts of neighbouring Algeria.

3. The Mongol nomads revered the wolf, the sole land-bound predator, to thrive for millennia on the steppes. It played a "paradoxical role in the people's [Mongols'] lives – as predator and prey, as defilers of herds ... and protectors of the grasslands' ecology ..." (Rong, *Wolf Totem*, viii).

4. Interestingly, the Mongol nomads who practised Lamaic Buddhism had a distinct aversion to water, believing that if they washed, they would be reincarnated as fish. In contrast, Muslim nomads valued the use of water, following strict Islamic precepts for personal hygiene.

Throughout many parts of the classical Islamic world, migrating nomads formed a ubiquitous feature of the landscape until very recent times. Over the centuries, both Muslim and non-Muslim travellers would mention them in their books, which described in detail the geography, history and ethnography of North Africa, the Middle East and Central Asia. However, it has only been in the twentieth century that we see a concerted effort by many governments in the region to curtail the free movement of nomadic groups within their jurisdiction.[5] This has not always been successful as nomads consider their independent lifestyle to be far superior to that of the sedentary dwellers of towns and cities. One must realise that the Arab, Berber, Bakhtiyari/Lur,[6] Turk or Mongol tribesmen made a conscious choice to lead the nomadic existence – this was not by accident.[7]

The Bedouin Arabs, the indigenous Arabic-speaking inhabitants of the Arabian deserts, became some of the very first Muslims. They are specifically mentioned in the Noble Qur'an in *Surah Tawbah*, where they are indistinguishable from any other people – some are more righteous than others, while others are more devout.

Throughout history, the terms 'Arab' and 'nomad' were often interchangeable. At other times, however, when the civilised sedentary Arab nation prospered, 'Arab' became synonymous with honour, justice and bravery. Conversely, in times of cultural deterioration, when waves of nomads poured out from the desert and overran urban centres, the term 'nomad' signified backwardness.[8]

---

5  Two noteworthy examples of this are the efforts of the nascent Soviet government in the late 1920s to forcibly settle the nomadic tribes of Central Asia and those of Reza Shah Pahlavi, who during the same period, tried to prevent the seasonal migrations of indigenous nomads throughout his native Persia.

6  The Bakhtiyari nomads, discussed in detail in Chapter 8 of this book, inhabit territory in western Iran to the south of 'Luristan', the ancestral home of the Lurs, with whom the Bakhtiyari nomads are related. The Lurs, as Indo-European ancestors of the Persian population, are considered to have dominated this part of Iran in the latter part of the first millennium BCE. Before this time, Luristan had been settled by the Elamites, as early as 3000 BCE.

7  Ure, *In Search of Nomads*, 155.

8  Al-Ansari, *Encounter of History and Modernity*, 83-85.

This book aims to inform the reader of the major nomadic groups that continue to inhabit ancestral parts of the Islamic world.[9] Their somewhat precarious existence in the early twenty-first century may seem an anachronism to some, but their survival (albeit in diminished numbers) into the digital age clearly indicates a hearty survival instinct and an outright rejection of the increasingly materialistic lifestyle of sedentary populations. And while all nomads are at one with nature, the Muslim nomad in particular is ideally suited to reflect on and react to the miraculous Signs of Allah by closely observing them in His environment: the spacious open skies, the star-filled heavens, snow-chilled streams, oceans of golden sand dunes, alluring desert oases with date groves, and lofty, snow-clad mountains.

**And the earth - We spread it out and cast therein firmly set mountains and made grow therein [something] of every beautiful kind. Giving insight and a reminder for every servant who turns [to Allah].** (50:7-8)

Nomads are by necessity spartan in their lifestyle. However, over many centuries they have developed an incredibly rich artistic culture of hand-crafted utilitarian objects. These sometimes exquisite works of art are invariably small so that they can be easily portable.

Some of the objects depicted in the chapters of this book are from the author's personal collection of nomadic trappings.

Islamic culture is rich and varied. The famed cities of Cordoba, Fez, Cairo, Damascus, Isfahan and Samarqand, to name but a few, all display the greatness and grandeur of the urbanised Islamic

---

9 The list of Muslim nomadic groups highlighted in this book is selective and by no means complete. For example, groups that have not been discussed include the nomadic Kurds, the Shahsavan (today known as the Ilsevan) of north-western Iran, the Afshars of the four corners of Iran, the Türkmen of north-eastern Iran, the Qaraqalpaqs of the Aral Sea delta, the Baluchis of eastern Iran, southern Afghanistan and Pakistan, and the Taimuri nomads of north-western Afghanistan.

world. The author maintains that nomads, sometimes living on the fringes, or in the shadow of these centres of civilisation, have also bequeathed to us a remarkable and equally stunning visual legacy – a heritage definitely worthy of our admiration and appreciation.

> "Rarely in human history have articles destined for hard use in a subsistence way of life reached such levels of complexity and beauty."[10]

This book is an introduction to Muslim nomads, those heterogeneous bands and tribes inhabiting Islamic lands, who migrate in search of greener pasture for their livestock. The political upheavals of the late twentieth century – some of which have continued into the twenty-first – have forever changed the nomadic lifestyle of millions throughout the Islamic world. These catastrophic events and their consequences will be addressed in subsequent chapters of this book.

10  Opie, *Tribal Rugs*, 6.

# 1

# WHO IS A NOMAD?

The word 'nomad' derives from the Ancient Greek word - *nomas* - 'one who roams about in search of pasture'.

The nomads who traverse the deserts, steppes and mountains of the Islamic world have always done so for a purpose; their wanderings are certainly not aimless, but very well-structured.

Over centuries, they have become adept at harvesting the scattered resources of the harsh environment of climatic extremes that they inhabit, and have developed a deep and natural attachment to the earth itself.

It is estimated that there are approximately 30-40 million nomads in the world today, and they fall into three distinctive groups. Foraging, in existence since the beginning of time, is the oldest form of nomadism. These hunter-gatherers were constantly on the move in search of food: game (deer, antelope, polar bear, walrus, seal and whale) that they could hunt and kill, and wild berries, nuts and fruit that they found along the way. Today, there are very few such nomads, the most well-known hunter-gatherers being the San Bushmen of the Kalahari Desert of southern Africa, and the tundra Inuit of Canada's Arctic North.

However, most traditional nomadic tribes today, many of whom are discussed at length in this book, are pastoralists. These communities breed herds of cattle and ruminants, such as goats and camels, and are in a perpetual search for new forage for their animals. Pastoral nomads such as, the Tuaregs of the Sahara Desert, may cross many hundreds or thousands of kilometres during their migrations, while the Qashqa'is of southern Iran may cover much less territory in their seasonal treks. Their wealth lies in their cattle, so once an area has been depleted of its herbage, the nomads are compelled to migrate elsewhere. Pastoralists struggle to maintain their independence in a world that increasingly favours sedentary existence.

'They have become adept at harvesting the scattered resources of climactic extremes that they inhabit'

The nomads seen in the industrialised West are peripatetic ones. These people are usually traders, craftsmen or farm workers, who move where their work takes them. The Gypsies or Roma (Romani) people are an itinerant group of nomads originating in northern India, from where they spread west into central Asia and Türkiye. They eventually arrived in Europe about a thousand years ago. The English word 'gypsy' derives from the Greek word for Egyptian, *Aigyptios*. The commonly-held belief in the Middle Ages was that these dark-skinned travellers had migrated from Egypt. Today, the peripatetic Roma are found in sizeable numbers in Türkiye and most European countries, as well as in North America and Brazil.

To what degree, if any, are pastoral nomads, (the focus of this book) reliant on the outside world? Is there any symbiotic relationship between nomadic and sedentary peoples? Anthropologist and historian, A.M. Khazanov, maintains that nomads could never exist totally independent from the sedentary agricultural and urban outside world with its diverse economic and social systems.[1] In the following chapter, we will examine what the great fourteenth century Arab historiographer, Ibn Khaldun, had to say about nomadism and its crucial role in social history.[2]

1  Khazanov, *Nomads and the Outside World*, 3.

2  'Abd al-Rahman ibn Muhammad ibn Khaldun al-Hadhrami (1332 Tunis -1406 Cairo) was an Arab Muslim social historian, who, over a long life closely observing the human condition, developed a unique theory of history that was eloquently laid out in his seminal work: the *Muqaddimah* ('Introduction'). The *Muqaddimah* presents Ibn Khaldun's theory that essentially discusses the causes of the rise and fall of empires throughout history.

# 2

# IBN KHALDUN
## AND BEDOUIN CIVILISATION

The French colonised North Africa (the 'Arab West') in the nineteenth century - Algeria in 1830, Tunisia in 1881, and Morocco in 1912. They theorised that, in the past, Morocco consisted of two distinct regions: the lands of state treasury, *bled al-makhzan*, and the lands of dissidence, *bled al-siba*.

While the *bled al-makhzan* comprised the urban centres controlled by the central government of the reigning sultan, the *bled al-siba* referred to the rural areas, whose inhabitants, mainly nomadic Berber tribes, had lived for centuries quite independent of any state control. The French, as colonial masters of North Africa, perpetuated the *makhzan/siba* division during their protectorate (1912-1956). Under the French, there developed a system of 'indirect rule' for the Berber tribal regions of *bled al-siba*.

The same *makhzan/siba* division forms an essential element in Ibn Khaldun's revolutionary 'theory of civilisation' that took him a lifetime to develop.

> Ibn Khaldun perceives history as a cycle in which rough, nomadic peoples, with high degrees of internal bonding and little material culture to lose, invade and take resources from sedentary and essentially urban civilisations. These urban civilisations have high levels of wealth and culture, but are self-indulgent, and lack both 'martial spirit' and the concomitant social solidarity. This is because those qualities have become unnecessary for survival in an urban environment, and also because it is almost impossible for the large number of different groups that compose a multicultural city to attain the same level of solidarity as a tribe linked by blood, shared custom and survival experiences. Thus the nomads conquer the cities and go on to be seduced by the pleasures of civilisation, and, in turn, lose their solidarity and come under attack by the next group of rough and vigorous outsiders – and the cycle begins again.[1]

1  Stone, *Ibn Khaldun and the Rise and Fall of Empires*.

> 'nomads… go on to be seduced by the pleasures of civilisation, and, in turn, lose their solidarity'

The nomad considers his own way of life superior to that of the town dweller or farmer. Free of material constraints, he feels himself more independent, manlier and braver, for where life is hard and necessities scarce, only constant struggle increases one's share. …Loyalty to family and tribe is essential for survival… The townsman, on the other hand, represents stability and continuity. By amassing excess wealth he is able to support art and education and the institutions of justice and religion.

But institutions grow old and decay, and the interaction between nomad and sedentary has historically played a major role in the renewal and strengthening of Islamic culture.[2]

Ibn Khaldun was a privileged observer of events occurring in his tribal world of fourteenth century North Africa. His remarkable ability to analyse the human condition is abundantly evident in the following passage from his *Muqaddimah*. What he says may sound rather obvious to us today, but one must remember that he was writing six centuries ago. For example, his detailed analysis of the relationship between the nomads of the desert and sedentary urban dwellers was totally revolutionary. It is specifically because of these insightful observations that he is considered by many to be the Father of Sociology.

It should be known that differences of condition among people are the result of the different ways in which they make their living. Social organisation enables them to co-operate toward that end and to start with the simple necessities of life, before they get to conveniences and luxuries.

2  Sabini, *The World of Islam Its Nomads, Its Cities*.

Some people adopt agriculture, the cultivation of vegetables and grain, (as their way of making a living). Others adopt animal husbandry, the use of sheep, cattle, goats, bees and silkworms, for breeding and for their products. Those who live by agriculture or animal husbandry cannot avoid the call of the desert, because it alone offers the wide fields, acres, pastures for animals, and other things that the settled areas do not offer. It is therefore necessary for them to restrict themselves to the desert. Their social organisation and co-operation for the needs of life and civilisation, such as food, shelter, and warmth, do not take them beyond the bare subsistence level, because of their inability (to provide) for anything beyond those (things). Subsequent improvement of their conditions and acquisition of more wealth and comfort than they need, cause them to rest and take it easy. Then, they co-operate for things beyond the (bare) necessities. They use more food and clothes, and take pride in them. They build large houses, and lay out towns and cities for protection. This is followed by an increase in comfort and ease, which leads to formation of the most developed luxury customs. They take the greatest pride in the preparation of food and a fine cuisine, in the use of varied splendid clothes of silk and brocade and other (fine materials), in the construction of ever higher buildings and towers, in elaborate furnishings for the buildings, and the most intensive cultivation of crafts in actuality. They build castles and mansions, provide them with running water, build their towers higher and higher, and compete in furnishing them (most elaborately). They differ in the quality of the clothes, the beds, the vessels, and the utensils they employ for their purposes. Here, now, (we have) sedentary people. 'Sedentary people' means the inhabitants of cities and countries, some of whom adopt the crafts as their way of making a living, while others adopt commerce. They earn

more and live more comfortably than Bedouins, because they live on a level beyond the level of (bare) necessity, and their way of making a living corresponds to their wealth.

It has thus become clear that Bedouins and sedentary people are natural groups which exist by necessity, as we have stated.[3]

Ibn Khaldun devotes an entire chapter (Chapter II) of the *Muqaddimah* to 'Bedouin Civilisation'. Here, he explains in detail his views regarding why the Bedouin society is in some ways superior to that of sedentary urban centres. The concept of '*asabiyyah* or strong 'group feelings' is what unites a nomadic tribe, and enables them to survive in the desert.[4] In fact, Ibn Khaldun "believed in a continuous cycle of progress and degeneration of human civilisation that was dependent on the strength or weakness of '*asabiyyah* within a population or nation… He stated that a ruler's true strength lay in the amount of '*asabiyyah* present amongst his people."[5] He further believed that the ultimate goal of the Bedouin, or desert nomad, is the urban lifestyle. The toughness of desert living and '*asabiyyah* prepares him for eventually conquering the 'soft' urban centres. Therefore, according to Ibn Khaldun, "Bedouins, thus, are the basis of, and prior to, cities and sedentary people."[6] The evidence for this is the fact that most inhabitants of any given city have nomadic ancestors originating in the surrounding desert.

3   Ibn Khaldun, *The Muqaddimah*, vol.1, 249-250.

4   '*Asabiyyah* may also relate to the close 'social solidarity' within a tribe or clan; Ibn Khaldun believed it to be the fundamental bond of human society; once weakened, the society can no longer survive – thus causing the rise and fall of dynasties and entire civilisations.

5   Nagy, *Ibn Khaldun The Maghribi Master of the Muqaddimah*, 59.

6   Ibn Khaldun, *The Muqaddimah*, vol.1, 252.

# 3

# BERBERS
## OF THE ATLAS MOUNTAINS

Berbers are considered to be the indigenous inhabitants of North Africa. No one knows the precise date of their arrival on the continent, but it was certainly many thousands of years ago. They are first identified as an independent people in an Ancient Egyptian text dating from 3000 BC.

Berber nomads call themselves *Amazight*, or 'free or noble men'. Unfortunately, they are known inthe West as Berbers, an offensive term originating in ancient times. The Ancient Greeks referred to all people not able to understand their language as *barbaroi*, or 'barbarians'. The Ancient Romans, not unlike the Greeks, used the term 'Berber' (from the Latin *barbarus*) to denote the incomprehensible foreign tribes they found inhabiting North Africa.

Throughout their long history, the Berbers were never a unified people. Today, approximately 25 million Berbers inhabit a broad swath of territory comprising many North African and Sahelian nations.[1] They can be found living along the Atlantic coast of Morocco, and as far east as the Siwa Oasis in western Egypt and from the banks of the Mediterranean Sea to the Niger River.

The Berbers, who account for sixty percent of Morocco's population, speak *Tamazigh*t ( ⵜⵓⵍⵓ⵿ⵣⵉⵢⵜ ) a very ancient language, which employs the equally ancient and unique *Tifinagh* script, derived from a Berber writing system in use for two millennia. A modified version of this ancient script is now employed in Moroccan elementary schools in teaching the Berber language, which became a mandatory subject in 2001.

A century ago the majority of Moroccan Berbers lived a rural semi-nomadic existence, making their semi-annual trek from the desert plains and mountain slopes to the grazing areas of the Middle and High Atlas Mountains. Today, the actual number of families making these migrations has fallen drastically. In 1988, for example, while several hundred families migrated, today only a dozen or so continue this tradition.[2] The migration route is not a haphazard one. To reach the grazing pastures, the nomads ply the same welltrodden sheep paths their ancestors have used for generations.

1   These countries are Morocco, Algeria, Tunisia, Libya, Egypt, Niger, Mali, Mauritania, and Burkina Faso.

2   Southam, *Morocco's Last Berbers on their 4,000-year-old Annual Migration- a Tradition that is now under Threat*.

When temperatures start to rise in spring, the Berber nomads proceed to migrate from the Saharan lowlands to the cooler, higher elevations of the Atlas Mountains where they erect their camel-hair tents.

When temperatures start to rise in spring, the nomads dismantle their camel-hair tents and proceed to migrate from the Saharan lowlands to the cooler, higher elevations of the Atlas Mountains. Moroccan Berbers all wear the practical *burnous*, a hooded woollen cloak; such a garment affords protection against the elements and can even act as a sleeping-bag, if necessary. They move all their animals - camels, horses, sheep and goats - like shepherds, regarding each animal as an individual that can be easily identified by markings or peculiar traits. Newborn animals too young to walk are placed atop saddle bags.

> If the horse is the essential element of transportation: the sheep is the essential element of almost everything else; mutton is for eating; ewe's milk for drinking and churning into pungent butter; sheepskins are for winter coats; wool is for making rugs for sale, and finally sheep-droppings are for compacting into bricks for fuel for the cooking and campfire.[3]

The Berbers transport all their belongings on pack animals. Many of the portable items are marvellous examples of nomadic weaving: tent bands, storage bags, coverings and carpets. These are all woven for a specific purpose using the wool of sheep or goats and camel hair. Nomads are also adept at using wild plants and flowers to make their colourful natural dyes. Today, these samples of nomadic art are coveted by collectors for their intrinsic beauty. Their makers couldn't have imagined that one day such utilitarian pieces would be so enthusiastically sought after by foreigners. Many scholars agree that the single greatest contribution made by nomads to Islamic art is their astonishing weaving expertise.

3  Ure, *In Search of Nomads*, 155.

The demands of a lifestyle and often precarious existence gave little time and few resources to devote to anything other than the needs of their families. In providing for these needs they created an impressive range of utilitarian textiles – pile rugs, flat woven covers and spreads; curtains and other trappings for the houses, tent or animals; bags, bands, belts and knitted items, such as socks – none of which required colour or design to be functional. The beauty with which they imbued these textiles no doubt gave them and their families as much pleasure as it gives us today. This artistic achievement is all the more impressive when seen in the context of their daily lives.[4]

Morocco gained full independence from France in 1956 after four decades of brutal colonial occupation. In 1912, the French permitted the establishment of a Spanish Protectorate – within Moroccan territory – along both the coast of the Mediterranean and its southern border. During the 1920s, the Berbers of the Rif Mountains were notorious for their resistance to Spanish colonial rule, in particular. The legendary Riffian Berber, Abd el-Krim (1882/83-1963), even established the short-lived Republic of the Rif (1921-1926). Similarly, in French-controlled Morocco, there was a fierce opposition to the occupation. A Saharan Berber, Cheik Mohamed Laghdaf (died 1960), is remembered for his refusal to submit to the foreign domination of his homeland. He struggled for decades against both occupying powers.

The tomb of Cheik Mohamed Laghdaf lies outside the southern Moroccan town of Tan-Tan. Since 1963, it has become the focus of an annual festival (called the *moussem* – Arabic for 'season'), which celebrates all facets of traditional nomadic culture. It has now become the largest gathering of nomadic tribes in all of

4 Wertime, *Sumak Bags of Northwest Iran and Transcaucasia*, 14.

All Berber nomads traditionally carried a very finely tooled leather satchel, which would be partially hidden beneath a burnous, a hooded woollen cloak. These bags were always made from the finest Moroccan leather, and would become more durable and attractive over time. Of the few material possessions Berber nomads prized, these were esteemed heirlooms passed down from father to son.

The Berbers transport all their belongings on pack animals. Many of the portable items are marvellous examples of nomadic weaving: tent bands, storage bags, coverings and carpets, such as this one. These are all woven for a specific purpose using the wool of sheep or goats and camel hair.

North Africa and is lauded by UNESCO as a "masterpiece of the oral and intangible heritage of humanity." Originally, the *moussem* was a religious festival which also offered the participants a chance to socialise and even sell their camels. But the gathering quickly expanded to include equestrian and camel-breeding competitions, poetry readings, story-telling, and even an opportunity to exchange herbal remedies. Today, approximately fifty nomadic tribes from across the Sahara (including Mauritania, Mali and Niger) attend the event. Foreign visitors are welcome and several tents offer thematic displays of tribal Berber life. Apart from the actual weaving of a traditional black goat hair tent, there are also displays of nomadic cooking and instruction in the Noble Qur'an.

Berbers learn Arabic as a second language. Nomadic families are sometimes fortunate to have an elder who can read and write Arabic and who might then be able to teach the rudiments of writing to the younger members. But nomads living near a desert or mountain village may be able to have their children study the Noble Qur'an at a local mosque. Throughout northwest Africa, a centuries-old tradition is followed in the teaching of the Noble Qur'an. Each child is given a thin, but sturdy, wooden writing tablet called a *lawha*, on which he or she practices copying verses from the Noble Qur'an, eventually memorising them. A simple reed pen and natural homemade ink are used. Once written, the text could then be examined and subsequently easily washed off. These wooden tablets are used for generations and over time develop a well-earned patina.

The Tan-Tan *moussem* highlights many important aspects of a rich nomadic lifestyle. The truth, however, is that nomadism in Morocco is seriously under threat with some predicting that within a generation, we will witness the last of the traditional Berber migrations. Climate change and deforestation are wreaking havoc on those who wish to continue the old lifestyle. In the past, a nomad's sheep, goats and camels could survive on juniper leaves if there was no grassland in a particular area. Now, the juniper-clad slopes of the Atlas Mountains are bare, forcing him to buy expensive fodder for his animals. Water is also an issue. Morocco, not unlike many countries in the region, is expected to experience more droughts in the future, which will surely have drastic effects on the few remaining nomads. Finally, the new generation is not so enamoured by the nomadic way of life. Those who do choose to pursue an education in one of Morocco's 'tent schools' must of course remain in one place. Education leaves them unprepared for the harsh life of their nomadic parents.

For millennia, environmentally friendly Berber nomads have survived amid the scorching Sahara sand dunes and chilly Atlas Mountains. But man's careless interference in the natural world is causing irreversible environmental damage mainly due to climate change (for example, global warming) and deforestation. All this may sadly cause the demise of the Berbers' migratory existence within our lifetimes.

Berber, or *Tamazight*, is a very ancient language. Although written, it has a wealth of proverbs that form part of a unique and extensive oral tradition. Among many wise sayings, the following proverb is especially memorable for its innate truth.

> "Courage is fear that has said its prayers."

Berbers of the Atlas Mountains    37

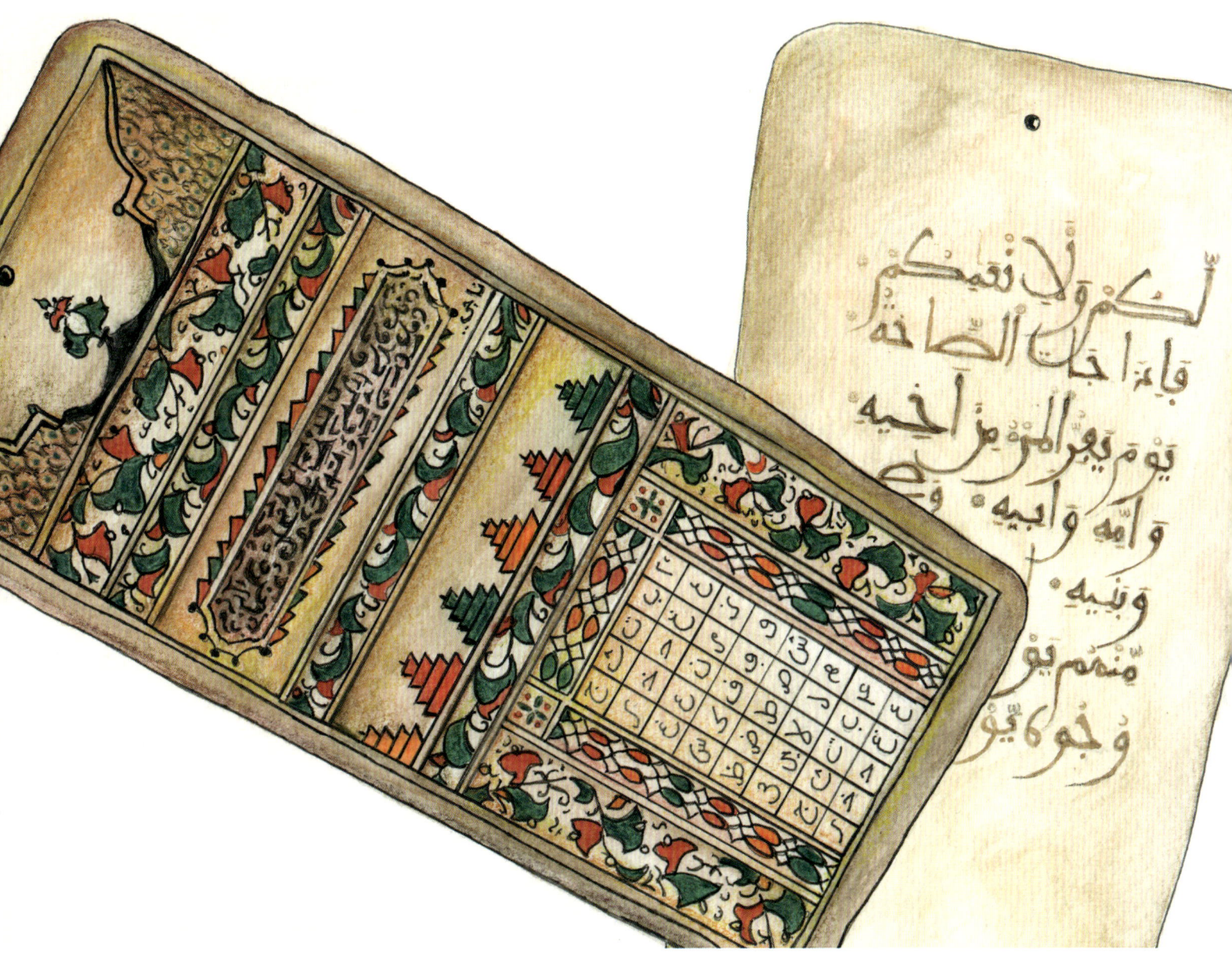

Throughout northwest Africa, a centuries-old tradition is followed in
the teaching of the Noble Qur'an. Each child is given a thin, but sturdy,
wooden writing tablet called a *lawha*, on which he or she practices copying
verses from the Noble Qur'an, eventually memorising them.

# 4

# TUAREGS
## OF THE SAHARA DESERT

The Tuaregs, better known in the West as the 'blue men' or 'veiled men of the Sahara', are pastoral nomads, whose economy is based principally on cattle-breeding and trading. They are genetically and linguistically related to the Berbers of North Africa.

While the term 'Tuareg' is of Arabic origin, the nomads all call themselves *Imuhagh* (related to Moroccan Berber *Imazighen*), which means 'freemen'. The Tuareg speak a Berber language called *Tamasheq* (once again related to the Moroccan Berber *Tamazight*). They herd their goats, sheep and camels across thousands of square kilometres of desert and rely solely on these animals for their sustenance. "Life to them [the Tuareg] was a bridge: one should cross over it, not build a house on it."[1] European explorers have commented on the fact that the Tuareg resembled their beasts of burden, the camel, in their ability to gorge and then fast for long periods of time – all this seemingly having no negative side effects.[2]

The origins of the Tuareg nomads predate history, but Herodotus, the classical Greek historian and 'father of history', mentions them in his writing as far back as the fifth century BCE. For centuries, the Tuaregs – now numbering more than one million – have dominated the caravan trading routes across the Sahara and the Sahel. Their caravans were renowned for transporting vast quantities of dates and other food stuffs, as well as enormous slabs of rock salt, which were often used as a currency in sub-Saharan regions of Africa.

As protection from the scorching sun and swirling sand, Tuareg men wear indigo-dyed turbans and robes, whose colour rubs off on the skin giving it a bluish hue. The veil, or turban, which covers the entire face with just the eyes and top of the nose exposed, is first worn at puberty and is considered the rite of passage to manhood.

1  Ure, *In Search of Nomads*, 11.
2  Ure, *In Search of Nomads*, 185-186.

Tuaregs of the Sahara Desert 41

As protection from the scorching sun and swirling sand, Tuareg men (like this one) wear indigo-dyed turbans and robes, whose colour rubs off on the skin giving it a bluish hue. The veil, or turban, which covers the entire face with just the eyes and top of the nose exposed, is first worn at puberty and is considered the rite of passage to manhood. This Tuareg nomad sits cross-legged on an elaborate fork horn saddle with his feet on the camel's neck.

Tuaregs generally live on the products of their goats and camels. Their homes are very practical, lightweight tents made of stretched goatskins that are stitched together and placed over a simple wooden frame. Utensils are very basic; the calabash gourd is grown as a vegetable, but when dried, can be used as durable bowls, jugs and ladles.

Early Muslim seafarers learned to use the celestial bodies in the night sky to successfully navigate across the Indian Ocean. The Tuaregs are true desert nomads who, over countless generations, have also refined their navigational skills by star gazing to help them safely traverse the oceans of sand. It is even said that some blind tribesmen, despite their apparent handicap, possess an innate, heightened sense of smell that enables them to lead a caravan of nomads through the merciless desert dunes.

Tuaregs generally live on the products of their goats and camels. Their homes are very practical lightweight tents made of stretched goatskins stitched together and placed over a simple wooden frame. Utensils are very basic; the calabash gourd is grown as a vegetable, but when dried can be used as durable bowls, jugs and ladles. Nomads, by preference and not out of necessity, are extremely self-sufficient, often preferring to drink camel milk for days or weeks on end, rather than seeking out wells.[3]

The traditional habitat of the Tuareg nomads covers the entire central Saharan region of Africa. Today, this area encompasses southern Algeria, south-eastern Libya, western Niger, northern Nigeria, northern Burkina Faso and southern Mali. It has become increasingly difficult for nomads to continue their migratory lifestyle in their ancestral homeland; this is due to the many political borders now in place as a result of the post-colonial division of Africa.

The Tuareg nomads have a complex social stratification made up of nobles, religious clerics and peasants who tend the desert oases in Tuareg territory, and bonded slaves. Although slavery has been banned throughout northwest Africa, the Tuaregs are

3  Ure, *In Search of Nomads*, 204.

reported to still continue the practice. Their society is patriarchal, in which the father is the supreme authority in the family, descent, and clan or tribe. However, it is also matrilineal, meaning that families trace their ancestry through the women, and not the men.

The recorded history of the Tuaregs dates back to the fourth or fifth centuries CE, to the time of legendary Tin Hinan. Venturing south from present-day Morocco, she eventually emerged in the Hoggar Mountains, the picturesque volcanic highlands of southern Algeria, where she settled and became the first queen of the Tuaregs. She is held in very high esteem and is today revered by the nomads as "the mother of us all".

The early Umayyad Muslims, who settled most parts of North Africa in the first Islamic century, invited all the indigenous peoples they encountered – Berbers and Tuaregs included – to embrace Islam. As the newly-converted Tuaregs travelled south into subSaharan Africa, they eagerly spread the message of Islam and were very effective in doing so.

Not unlike their Moroccan Berber brethren, the Tuaregs also fiercely resisted the French colonial occupation of their desert homeland. Although it has taken a long time in coming, there is today an active Tuareg ethnic revival. This stems from the rise in the 1990s of Berberism, a fervent Berber cultural movement originating in Algeria and Morocco. This same movement spawned several attempts for Tuareg independence in Niger and Mali. Unfortunately, in their struggle for some nominal autonomy from Saharan governments, some Tuaregs have partnered – to their detriment – with extremist Islamic groups such as ISIS and Boko Haram.

The recorded history of the Tuaregs dates back to the fourth or fifth centuries CE, to the time of legendary Tin Hinan. Venturing south from present day Morocco, she eventually emerged in the Hoggar Mountains, the picturesque volcanic highlands of southern Algeria, where she settled and became the first queen of the Tuaregs. This portion of the central Sahara, the Hoggar massif, is known as the 'land of the Tuaregs'.

The Tuaregs have ridden camels for as long as they have inhabited the Sahara. Their fork horn saddles are elaborately crafted out of wood and metal (copper, silver and brass ornaments) and have the distinction of being one of the most difficult saddles to keep your balance on. Balancing is made even more difficult as a rider's feet must rest on the camel's neck. Unlike other saddles, these are placed in front of the camel's hump, over its strong shoulders.

If destructive political alliances of the twenty-first century do not hasten the demise of the Tuareg nation, environmental factors may. Despite the aridity of their ancestral habitat, the Tuaregs have managed to survive. Sadly, their days as free, independent nomads may be numbered. The Saharan aquifers (underground reservoirs of 'fossil water') are being depleted at an alarming rate in Niger, for example. This is mainly due to the government's decision to exploit the known deposits of uranium – an extremely water-intensive process. Water scarcity has led directly to the desiccation of Tuareg grazing lands. Some young Tuaregs consider their traditional culture outdated and now wish to adopt a more sedentary lifestyle, even preferring Arabic to their ancient native tongue.

For many of the world's indigenous peoples, the past century has been a very painful period of forced transition into modernity. Can the overwhelming tide towards materialism and uniformity – the hallmarks of globalisation – ever be reversed? Once aboriginal peoples renounce their traditional lifestyles and adopt a sedentary existence in towns and cities, the first fatality of such a move is the rapid extinction of their native languages. What indigenous peoples, including many of the nomadic groups discussed in this book, know about the natural world often exceeds what is known by the scientific community.[4] The loss of these languages signals the loss of hundreds, if not thousands, of years' worth of accumulated knowledge that can never be retrieved.

> Much of what science does not yet know about the environment is known by speakers of endangered languages. Much (if not most) of humankind's accumulated knowledge of the natural world is encoded solely

[4] Harrison, *The Last Speakers*, 19.

in languages that have never been written down or documented and are now facing extinction.

> With language extinction, we lose human knowledge about the natural world... Languages uniquely encode in their grammars and lexicons specific information about topography, endemic species, and other environmental factors such as weather patterns and vegetation cycles... Most of this information is packaged such that it cannot be directly translated. Such knowledge erodes or dissipates when a community shifts over to speaking a global language...[5]

The Tuaregs as nomads by preference and not necessity are at a crossroad. Only time will determine their survival in the fast-changing twenty-first century. The following two Tuareg proverbs give one hope that their unique form of nomadism will not succumb to materialism.

> *"Better to walk with not knowing where than to sit doing nothing."*

> *"That which you do not need will kill you."*

---

[5] Harrison, *The Last Speakers*, 243-244.

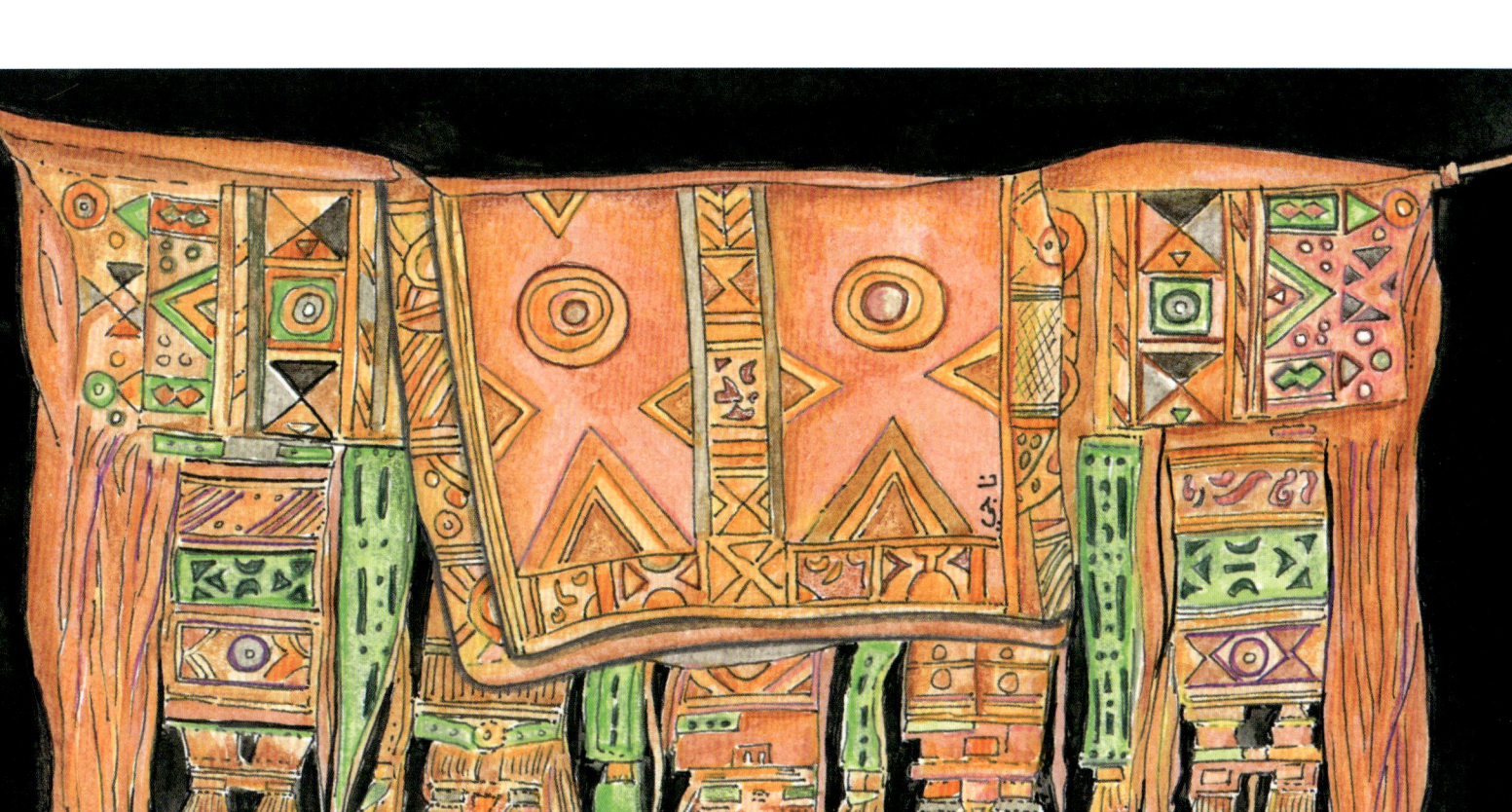

Like their Moroccan relatives, the Atlas Berbers, the Tuareg nomads make very distinctive
and attractive leather bags. This is a remarkable example of a Tuareg woman's travelling
bag made of subtle goat skin. The decorative patterns are ancient and symbolic.

# 5

# YÖRÜKLER
## OF THE TAURUS MOUNTAINS

The pastoral nomads of Türkiye, the *Yörükler*, are thought to have originally migrated westward from Central Asia in the eleventh century.[1] They first entered Anatolia (the highlands of central Türkiye) with the waves of Seljuq migrants immediately after the Battle of Malazgirt in 1071.[2]

---

1  The term derives from the Turkish verb *yürümek* ('to walk'); *Yörükler*, therefore, means 'those who walk', i.e., the 'walkers' or 'wanderers'.

2  The decisive battle took place on August 26, 1071 on the plains of Malazgirt in eastern Türkiye. The Turkish Seljuq army led by Alp Arslan defeated the Greek Byzantine Emperor Romanus Diogenes IV. This proved to be one of the most important battles in the history of Western Asia. The Sejuqs were thus able to extend the frontiers of Islam into the very heart of Byzantine Christian territory. In an extremely short period of time, vast tracts of former Greek-speaking land became forever Turkish. (Nagy, *The Book of Islamic Dynasties*, 47, 49.)

Some of these nomads eventually settled down and populated the four corners of Anatolia. Others, however, chose to continue their nomadic lifestyle.[3] One such group inhabited the Taurus (*Toros*) Mountains, which skirt the coastline of southern Türkiye. Here they could make their semi-annual migrations to their summer and winter quarters. Descendants of these nomads continue to do so even today. Turkish culture is a fusion of the traditions of the many indigenous peoples of Anatolia, along with the Turkic nomads who entered the area centuries ago.

The Yörükler of Türkiye are sometimes mistaken for indigenous gypsies. While the latter are itinerant travellers, who may traverse great lengths of the country, Yörükler tend to travel along shorter more traditional routes from the mild coastal lowlands (*kışlık*) to the lush pastures of the mountain highlands (*yaylak* or *yayla*).

The Yörükler of the Taurus Mountains can take between two to four weeks to complete their seasonal migrations. In preparation for the move in late spring, everyone works: women sew clothes and bake large batches of long-lasting çörek (a type of unsweetened shortbread) for the trek, and men repair the gear for the pack animals (reins, harnesses and saddles). Originally, Yörükler used camels for transportation. Today, their possessions are most likely transported by donkey or mule.[4] The nomads' diet is a basic but nutritious one consisting of *bulgur* (cracked wheat), wild herbs, fresh butter, cream, cheese and yogurt all made from sheep and goat's milk.

3  For example, it is said that until the end of the nineteenth century, hundreds of thousands of Yörükler migrated throughout Ottoman territory in the Balkans and Anatolia.

4  Camels can still be seen, but horses aren't sure-footed enough to make the treks into the Taurus Mountains.

The Yörükler inhabit the Taurus (*Toros*) Mountains, which skirt the coastline of southern Türkiye. Yörükler tend to travel along shorter more traditional routes from the mild coastal lowlands to the lush pastures of the mountain highlands. Yörükler are humble folk in harmony with their natural habitat; in many ways their lives are extremely environmentally friendly.

Yörük women play a very important role in their society. The oldest woman in the family looks after the tent – the family's home. Women weave durable panels of black goat's hair that are sewn together to make the outer shell of the tent. The inner shell is lined with thick felt, which offers natural insulation and waterproofing. Within the tent, women spin and weave, light the fires, cook the meals, bake unleavened bread and seek shelter when giving birth.

Yörükler are humble folk in harmony with their natural habitat. In many ways their lives are extremely environmentally friendly, leaving a low carbon footprint. Yörükler, like the Native American Indians of the Great Plains, are very careful to avoid any wastage and to make use of all parts of any slaughtered animal, including its horns and entrails. As Muslims, they offer Friday prayers and the two 'eid prayers in the nearest village mosque.[5] When up in the mountain pastures, however, they improvise and construct a simple open-air *namazgâh* ('place for praying'). This consists of hand-woven *kilims* (flat weave rugs) placed on a flat section of ground, which is then encircled by a barrier of stones.

The *kilim* is considered to be the oldest form of nomadic weaving. Amazingly, some ancient designs in Neolithic wall paintings from the prehistoric site of Çatal Hüyük (6350-5400 BCE) resemble patterns on some Anatolian *kilims*. Some scholars, however, believe this to be coincidental and contend that the *kilim* was first introduced into Anatolia by the Turkish nomads who arrived from Central Asia. What is undeniable is the fact that many *kilim* designs, centuries old, have survived and can be seen in the *kilims* of today.

5  The Taurus Yörükler are a pious folk. At times, water can be scarce, but in their tents, there is always a ewer of warm water available for the five daily ritual ablutions.

In a society that had no access to other artistic media, tribal weavers used the *kilim* designs as an expression of their own culture and tribal identity. A *kilim* pattern belonging to one tribe was never reproduced or copied by a weaver belonging to another tribe. The designs were strongly symbolic.[6]

The art and craft of weaving is endemic in all nomadic societies in the Islamic world. For generations, traditional weaving techniques and designs have been passed down from mothers and grandmothers to daughters and granddaughters.[7] Besides the laborious task of weaving, several additional skills are required before a carpet or storage bag can be produced. For example, wool sheared from goats or sheep first has to be sorted, combed, spun and, if necessary, dyed. Nomads instinctively know that the best quality wool comes from the neck and shoulder areas of a sheep. Sheep that have been able to graze all summer long in green pastures can be sheared once a year and produce a good quality fleece. The poorer quality grades are used for making felt.

Although the wool of goats, sheep and camels can produce an astonishing variety of beautiful natural hues,[8] nomads have for centuries also used natural plant-based dyes to add extra colour to their weavings.[9] Almost any part of the plant – flowers, leaves, stems, roots – could produce a dye. The indigo plant supplied a rich dark blue, while the dried roots of the madder plant generated a remarkable variety of hues: pink, rose, apricot, scarlet, rust red, and purplish-brown. Yellow could be obtained from many wild indigenous plants of the Central Asian steppes and Anatolia. Saffron crocus stamens and pomegranate rinds are two good examples. For other colours, two or more dyes were mixed; for example, indigo plus a yellow dye produced green.

6  MacDonald, *Tribal Rugs Treasures of the Black Tent*, 59.

7  In the tumulus grave sites in Central Asia, weaving and spinning tools dating back thousands of years have been discovered by archaeologists.

8  Sheep's wool, for example, is naturally produced in colours ranging from white, to grey, silver, brown, red and black.

9  The world's oldest pile carpet, found in a frozen Sythian (Eurasian nomads) burial mound in the Pazyryk Valley of the Altai Mountains of southern Siberia, dates from the fifth century BCE. It was woven using the natural colour dyes of red and blue.

Yörükler offer Friday prayers and the two 'eid prayers in the nearest village mosque. When up in the mountain pastures, however, they improvise and construct a simple open-air *namazgâh* ('place for praying'). This consists of hand-woven *kilims* (flat weave rugs) placed on a flat section of ground, which is then encircled by a barrier of stones.

Each tribe of nomads had its own dyes and *kilim* patterns, which were handed down over the generations. Unfortunately, today most nomads no longer use vegetable dyes in their weaving, but prefer synthetic ones, which first became available in the late nineteenth century.

The nomads of present-day Türkiye are renowned for their expertise in the production of *kilims*, which they weave on horizontal looms. This type of loom is preferred by the Yörükler as it can be both easily erected and dismantled. The width of such a loom is of course limited. Many larger *kilims* are therefore created by sewing two narrow *kilims* together. The Yörükler of the Taurus Mountains of southern Türkiye have a long and illustrious history of *kilim*-weaving.

Nomadism is sadly on the wane in Türkiye. For centuries, governments have tried to enforce the settlement and relocation of Türkiye's nomadic populations. Kemal Atatürk (1881-1938), the founder of the Turkish Republic, gave the indigenous nomads of his country a special status as he considered them the true standard-bearers of traditional Turkish culture.[10] However, after the death of Atatürk, there was a greater effort on behalf of the government to sedentarise all nomads. Generous incentives encouraging settlement have been offered by other governments. Water scarcity due to deep drilling and the reduction in the amount of available summer grazing because of urbanisation have led many Yörükler to yield to government demands. Still there are approximately four hundred Yörükler families who continue to migrate as before.[11] Their seductive caravans of tinkling camel bells can be seen in spring and early autumn as the nomads move from the lowlands to the highlands.

10  Hull, *Kilim*, 108.
11  Dr. Fatih Uslu (sociologist, Akdeniz University, Antalya, Türkiye) in discussion with author, June 2016.

◂ The Yörükler of the Taurus Mountains of southern Türkiye have a long and illustrious history of *kilim*-weaving. A variety of utilitarian items are woven, including storage bags, saddle bags, tent straps (*malbands*) and floor coverings. Each tribe of nomads has its own dyes and *kilim* patterns, which are handed down over the generations.

A newer generation is discovering its ancestral nomadic 'roots' and attempting to revitalise forgotten weaving traditions. But long gone are the days of the mass migration in the spring of thousands of Yörükler with their animals and belongings. Today, there is a sincere effort spearheaded by a group of Turkish academics to encourage nomads to preserve their sustainable lifestyle in the belief that in so doing we can all benefit. With increased awareness, it is hoped the Turkish population will reconnect with their ancestral roots and will be eager to preserve the country's nomadic heritage.[12] Some nomads now comfortably settled in towns and villages fondly recall the days when they wandered freely in the open spaces of Türkiye's Anatolian plateau. And they will quickly recall one of their ancient proverbs:[13]

*Yörük gün biterse,* çadır *kurup oturur.*
Ömrü *biterse, oldugu yerde gömülür.*
When the day ends, a Yörük erects himself a tent
and stays there.
When his life ends, he is buried on the spot where he has died.

12 To this end, the author has suggested to some academics that the government should proclaim a "Yörük Day", an annual event that will celebrate Turkish nomadism in all its many facets.

13 There are literally hundreds of unique proverbs indigenous to the Taurus *Yörükler*. Although these are for the most part transmitted in modern Turkish, they are still distinctly nomadic and sound totally new and original to an urban Turk. By careful examination, one can discover their antiquity. For example, the following saying clearly shows that the Taurus Yörükler have inhabited the Taurus mountains for a very long time and are familiar with its fauna. Today, there are no lions found in the wild in any part of Türkiye, yet the proverb mentions it. *Yatan aslandan, gezen tilki yegindir.* "A wandering fox is better than a sleeping lion."

The nomads of present-day Türkiye are renowned for their expertise in the production of *kilims*, which they weave on horizontal looms. This type of loom is preferred by the Yörükler as it can be both easily erected and dismantled. The width of such a loom is of course limited. Mothers teach their daughters the secrets of weaving at an early age by providing them with a smaller, modified loom.

# 6

# BEDOUINS
## OF ARABIA

Apart from the very few small urban centres of Yathrib (Madinah), Makkah and the cities of Yemen, most of the population of pre-Islamic Arabia was rural. The Arabian Desert was inhabited by pastoral nomads, or Bedouins, who shared a common culture of goat and camel herding.[1]

---

1  The term 'Bedouin' is derived from the Arabic word for 'wanderer', or 'nomad'. The word also derives from the same Arabic root as the word for 'beginning'. Urbanised Arabs believed that their ancestors were desert Bedouins. Ibn Khaldun, the great Arab social historian, clarified this point in his *Muqaddimah* (See Chapter 3 of this book). In the Noble Qur'an, the term 'Bedouin' is not found; the Prophet Muhammad ﷺ is referred to as an 'Arab (Surah Fussilat: 44). Today, the term is inextricably linked with the Arab nation.

Arabs believed that all sedentary Arabs were descended from desert nomads.[2] According to Ibn Khaldun, the desert-dwelling Bedouins have always had a relationship with urban centres. "Evidence for the fact that Bedouins are the basis of and prior to, sedentary people is justified by investigating the inhabitants of any given city. We shall find that most of its inhabitants originated among Bedouins dwelling in the country and villages of the vicinity."[3]

The pre-Islamic Arabian Bedouins wandered in search of pasture for their animals and in doing so became experts in traversing the desert wastes. As another source of income, they would safely transport goods and people in caravans, which followed ancestral routes known only to them.[4] The lucrative trade in expensive incense was dependent on Bedouins transporting the commodity from the cities of southern Yemen along caravan routes, which passed through Makkah, Yathrib and eventually reached the trading emporium of Petra, in present-day Jordan.[5]

> The symbiosis between the sedentary and the nomads is nowhere more apparent than in the ancient trade patterns of the Arabian Peninsula. ... The Southern Arabs transported gold, precious stones, silk and spices from India and the Far East across the seas to their home ports and then enlisted the services of their northern neighbours, the Bedouins, to transport the goods by camel caravan across the deserts to the Mediterranean world.[6]

---

2 It is noteworthy that the Islamic state first established by the Umayyads (stretching from east and west from its capital of Damascus) was consolidated by tens of thousands of Bedouin nomads who migrated to the far corners of the new Islam. In time, this Arab-controlled state was in turn conquered by successive waves of nomadic Turks, Mongols and Berbers.

3 Ibn Khaldun, *The Muqaddimah*, vol.1, 253.

4 Such commercial ventures continued into the Islamic era. The Prophet Muhammad ﷺ himself, along with his uncle Abu Talib, followed the same ancient routes and led trading missions to Syria.

5 Petra, the secluded capital city of the Nabataean Arabs, lay close to the ancient Arabian caravan routes. It was an extraordinary city carved from rose coloured sandstone mountains; by the third and second centuries BCE, it had become wealthy and powerful from the many caravans that passed through its territory.

6 Sabini, *The World of Islam Its Nomads, Its Cities*.

Bedouins of Arabia 63

In ancient times, the lucrative trade in expensive incense was dependent on Bedouins transporting the commodity from the cities of southern Yemen along caravan routes, which passed through Makkah, Yathrib and eventually reached the trading emporium of Petra, founded by Nabataean nomads, in present-day Jordan. Today, descendants of these nomads still inhabit the Petra valley.

Pre-Islamic Arabian society was illiterate except for a small number of scribes who were able to read and write Arabic using the ancient *musnad* alphabet.[7] However, all Arabs prided themselves on their ability to speak a unique and very rich language, "the language of the *dhad*".[8] They admired their gifted poets, the two most famous being Imru' al-Qais and 'Antara. During the annual tribal poetry competition held as part of the Ukaz Fair (not far from Taif), winning poems were 'suspended' (*al-mu'allaqat*) either on or in the Ka'bah in Makkah. The Bedouins of the Arabian Peninsula believe they are descended from one of two groups. Yemenis are considered descendants of two legendary figures, Adnan and Qahtan, while the Qaysis tribe of north central Arabia claim Ismail ﷺ as their ancestor. Over the centuries, nomadic family groups formed clans which in turn developed into larger tribal units. In Arabia's harsh desert environment, close co-operation amongst tribal members would ensure survival and protection from hostile enemies.

The Arabian Peninsula was the homeland of the Bedouins, but during the seventeenth and eighteenth centuries, various tribes migrated north and settled in present-day Syria. Despite the political instability and chaos resulting from wars and coups d'état, the Bedouins and their simple pastoral lifestyle survived intact until the middle of the twentieth century. However, a severe three-year drought beginning in 1958 effectively caused many Bedouins in Syria to abandon their camel-herding lifestyle for a more sedentary one. At that time, the ruling Ba'ath Party was also intent on eliminating all traces of tribalism within the borders of Syria. But since 1970, the al-Assad dynasty has attempted to strengthen its support base by introducing reforms in favour of the Bedouins.

7  The ancient South Arabian languages (including the languages of the kingdoms of Saba, Himyar, and Qataban etc.) were written in the unique *musnad* alphabet. *Musnad* continued to be used until the sixth century CE; it was eventually replaced by an early version of today's Arabic alphabet.

8  Arabs believed that their language was unique in haviing this sound, represented by the Arabic letter *dhad* (ض ).

The Syrian uprising, which started with the government's repression of peaceful protests in Deraa in March 2011, has understandably caused great division within the country. While most Bedouins support the opposition and continue to defend themselves against the al-Assad forces, some have had to flee their homeland and have become refugees. According to Dawn Chatty, Professor of Anthropology and Forced Migration at the University of Oxford, the key to a post-conflict unified Syria is to make use of the Bedouins' invaluable, deep-seated networks and social links to Syria's disparate ethnic population.

> Bedouin tribes have been virtually left out of contemporary Arab politics, often according to regime dictates. Successive Syrian governments have sought to officially delegitimise the country's Bedouin tribes, ignore them altogether, or co-opt them for regime gain. But none of those efforts ever made the tribes disappear. In fact, in recent years, tribal self-identification in Syria has only increased, and tribes' involvement in the Syrian uprising signals that they should not be underestimated as the country's future unfolds.[9]

Jordan, another artificially-created state recognised by the League of Nations in 1922, has a population of which a third or more are Bedouins. Before the creation of the state of Israel state in 1948, there were sizable numbers of Bedouins inhabiting the Negev Desert.[10] Their population decreased dramatically as Jewish settlers began to occupy their land and state policies ordered their transfer to government-approved 'townships' in the Negev. This forced settlement of Palestinian nomads has had disastrous results. A once free and independent people, who had

9  Chatty, *Syria's Bedouin Enter the Fray*.

10 These nomads roaming the deserts with their goats, sheep and camels lived as they had done for generations. They harmed neither man nor their environment; they lived off the grid leaving a carbon-free footprint.

easy access to watering holes throughout the desert, are now immobile and totally dependent on a hostile government for their survival.

The plight of the Palestinians – urban, rural and nomadic – is an ongoing tragedy the world has been painfully witnessing for well over half a century. One source of their suffering is the distortion of historical reality. The state propagated the myth that Palestine was devoid of Palestinians until they began to develop the country. The mass emigration of Ashkenazi (Central and Eastern European) Jewry to Palestine was intensified by the spread of such disinformation. A popular fallacious slogan during the nineteenth and early twentieth centuries stated: "A land with no people for a people with no land."

Palestinians – pagan, Christian and Muslim – have of course been indigenous to the region for thousands of years. Their presence is attested by Roman and Ottoman records.

The Bedouins are renowned for the purity of their speech. Even the Prophet Muhammad ﷺ was sent as a child into the desert to live amongst the Bedouins, from whom he learned how to speak the most unadulterated form of Arabic. They are equally renowned for their extremely generous hospitality; for a Bedouin, being called a miser is the utmost of insults. Providing shelter and food to a stranger takes precedence over one's own comfort. This same hospitality also applies to one's enemies.[11] Etiquette demanded that a stranger first tether his camel and make it kneel before approaching a Bedouin's tent from the front.

---

11  According to the traditional 'salt bond', a Bedouin's enemy was protected for three days after leaving the host's tent. Three days was believed to be the amount of time food and drink remained in the stomach.

> 'for a Bedouin ... Providing shelter and food to a stranger takes precedence over one's own comfort.'

Wilfred Thesiger (1910-2003), the British explorer and travel writer, had many first-hand experiences of Bedouin hospitality during his crossing of Arabia's Rub' al-Khali Desert. In his famous book, *Arabian Sands*, he related the story of an extraordinarily generous Bedouin who became a pauper by slaughtering his last camel to provide hospitality for his guests. Thesiger reminded his readers that the desert had been dominated by the Bedouins for a period longer than all the ancient civilizations. These nomads were totally at home in the desert environment, no matter how inhospitable it might be. Thesiger believed in the adage: the harder the life, the better the man. All creature comforts were to be foregone in the pursuit of this noble cause. And Bedouins' trust was in their Maker. "One paradox was the Bedouin attitude to weather. Sandstorms and other natural occurrences were of overriding importance for their effects of travel and safety, but the Bedouin would never speculate about or attempt to forecast the weather: it was the prerogative of Allah and to anticipate His will was a form of blasphemy."[12]

Coffee drinking is a pastime enjoyed by all Bedouins. When the nomads are not migrating, they might prepare the beverage several times a day. If a guest arrives, he is honoured by having the coffee making ritual performed in front of him. Coffee beans and cardamom are carefully stored in small utility bags. First, the required amount of beans are roasted and stirred in a long-handled iron pan. The hot beans are then transferred to a wooden tray for cooling. The roasted beans are then emptied into a mortar, where they are ground into a powder. An elaborately designed hourglass-shaped *dallah*, or coffee pot, of water is brought to the boil over a fire fueled by camel dung. The boiling water is transferred to a second *dallah* containing the ground

---

12  Ure, *In Search of Nomads*, 123.

coffee beans. This is all brought to the boil again. Finally, ground cardamom is added once the coffee has settled to the bottom of the *dallah*. This mixture is poured into a second brightly-polished *dallah* and then served to the guests in tiny, handle-less china cups. Custom dictates that one drink only three cups and then decline more by shaking the empty cup with quick movements of the wrist.

Migrating Bedouins transport all their worldly belongings by camel. Their tents therefore have to be portable, durable and easy to erect. The black goat hair tents (*bayt al-sha'ar* or 'house of hair'), seen wherever the Bedouins roam, is a remarkably practical and comfortable dwelling.

> The Bedouin tent is a marvellous adaptation of simple materials to demanding requirements. The modular goat hair tent has been perfected over the centuries by its makers. It is customary for women to weave long bands of heavy fabric that are stitched together to make the rectangular shelter. The size of the tent often depends on the carrying capacity of the family camel. A typical tent measures 7 metres by 13 metres and weighs approximately 200 kilos.[13]

Despite the size, guests were welcome in any nomad's tent. Protection was automatically given to anyone who pitched their tent next to a fellow Bedouin. He became a 'tent neighbour' and could ask for permission to stay with his neighbour and even join in their tribal migration.

13  Nagy, *Green Muslims*, 33.

Coffee drinking is a pastime enjoyed by all Bedouins. When the nomads are not migrating, they might prepare the beverage several times a day. If a guest arrives, he is honoured by having the coffee making ritual performed in front of him. The coffee is served in an elaborately designed hourglass-shaped *dallah*, or coffee pot.

It is typical for Bedouins to erect their tents around wells in desert oases owned by the tribe. But once winter rains fall, the nomads will quickly take their camels and goats into the open desert seeking out the vegetation that sprouts up around rock pools and other shallow areas where the rain water might have collected.

The Bedouin tent is the ideal sustainable dwelling for these desert people.

> The goat hair is woven loosely to allow for ventilation. Further natural air conditioning is possible by rolling up the side and back panels of the tent. Its dark colour creates shade which insulates against heat. If it rains, the tent will remain waterproof as the goat hair fabric contracts tightly when wet. Goat hair also retains heat on cold desert nights. The open end of the tent is always carefully oriented away from the prevailing wind.[14]

Bedouin women play a crucial role in the pitching and dismantling of the tent. Once erected, their personal possessions, kitchen utensils and provisions would be placed in the centre of the tent. Curtains made from the same woven goat hair panels divide the tent into two or three sections. Guests are welcomed in the men's quarters; a hearth is situated nearby and furnished with rugs, cushions and coffee-making equipment; and each tent also includes a kitchen and family/sleeping area. Bedouin women also care for their children, cook meals, gather fuel, weave rugs and other utility items, and might even help in the herding of the animals. In the past, the female members of a wealthy Bedouin family would ride in a covered-in camel litter (*hawdaj*), which was

14  Nagy, *Green Muslims*, 35.

Bedouins of Arabia 71

Migrating Bedouins transport all their worldly belongings by camel. Their tents therefore have to be portable, durable and easy to erect. The black goat hair tents (*bayt al-sha'ar* or 'house of hair') seen wherever the Bedouins roam, is a remarkably practical and comfortable dwelling – "a marvellous adaptation of simple materials to demanding requirements."

placed atop a camel's back. The litter was covered with a black cloth for privacy and protection from the sun.

As we have seen in previous chapters of this book, nomads have become proficient in the art of weaving. Relying solely on their animals for basic raw materials - sheep's wool, goat and camel hair - they are able to weave the essential tent cloth and all other items needed for daily use: utility bags, saddle bags, blankets, tent bands, and dividing curtains. The Bedouins of Arabia, not unlike their nomadic brethren of North Africa and the Sahara, are adept as weavers. Despite the availability of imported fabrics, weaving is still practised by some Bedouin tribes. The simple, practical, portable and horizontal ground loom is preferred by Bedouins who continue to migrate from one region to another. They use the wool from sheep and the hair from goats and camels to weave items of exceptional quality and beauty. Although gaudy synthetic dyes are readily available and, indeed widely used by Bedouin weavers, undyed wool (in varying shades of brown, grey and black) continues to be used by some as are natural dyes extracted from pomegranate rinds, walnut husks and onion skins, for example.

Since the middle of the twentieth century, many Bedouins in Saudi Arabia have chosen the sedentary lifestyle of the fast-changing urban centres throughout the country. Undoubtedly, as Bedouins say farewell to their ancient nomadic ways, a wealth of valuable traditional skills have become endangered. However, the Al Murrah of the Rub' al-Khali, are one tribe that has continued to preserve some of these skills. Over countless generations, the Al Murrah Bedouins have traversed this inhospitable habitat using only the stars and their sharpened senses to guide them.

So fine-tuned is their tracking expertise, an Al Murrah Bedouin's testimony is accepted in a Saudi court of law – as if it were a set of fingerprints. A Saudi saying alludes to the innate abilities of these gifted nomads: *Fi alsama' bariqiyah, fi al-'ardh Murriyah* ("In the sky the telegraph, on the ground the Al Murrah").

Bedouins regard the camel as a "gift from Allah" because of its importance in every aspect of their lives. Not only is the animal able to transport a family's tent on its back, it is the Bedouin's main source of staple foods: meat, milk and other dairy products. The camel has been called the "ship of the desert". This is because it is able to traverse vast distances without being dependent on food and water supplies. It is the only animal of Allah's creation with feet especially adapted for naturally walking over the hot desert sands. Like ships sailing on an open sea, the swaying motion of camels on the move over an open desert can induce a sort of sea sickness in their riders.

Unprecedented droughts, government settlement policies (in Tunisia, Libya, Egypt, Palestine, Jordan and Iraq) and the lure of a more comfortable lifestyle have induced many Bedouins to abandon their ancestral life as stateless nomads. Today, for those settled in the noisy, congested towns and cities of the Middle East, sadly the only reminder of their parents' or grandparents' nomadic past are the stuffy canvas tents (often artificially air conditioned) that have become a permanent fixture within the tiny, concrete-walled gardens of their villas. An old Bedouin will be saddened by the sight of his children and grandchildren entrapped by today's rampant consumerism, who may actually prefer to live within concrete walls. He will remind them that things were quite different in the past. For him, luxury

In the past, the female members of a wealthy Bedouin family would ride in a covered-in camel litter (*hawdaj*), which was placed atop a camel's back. The litter was covered with a black cloth for privacy and protection from the sun.

automobiles, chandeliers, state-of-the-art artificial air conditioning and widescreen plasma TVs have little worth.

*"When you sleep in a house, your thoughts are as high as the ceiling, but when you sleep outside, they are as high as the stars!"*

*"My treasures do not clink or glitter; they gleam in the sun and neigh in the night."*

# 7

# BAKHTIYARIS
## OF WESTERN IRAN

In his *Muqaddimah*, Ibn Khaldun, the great North African social historian, defined nomads as "people who make their living by rearing animals ... and are obliged to move and roam in search of pastures ... and water."[1]

1  Ibn Khaldun, *The Muqaddimah*, vol.1, 253.

He clearly identified '*asabiyyah* (strong communal pride and social cohesion) as the fundamental bond in human society, the cement that indelibly unites a tribe having a shared ancestry.² Therefore, according to Ibn Khaldun, nomads "unlike sedentary peoples attach more importance to descent than to domicile."³

Today, Iran hosts many tribes, some of whom still practise the traditional pastoral nomadism of their forefathers. The Bakhtiyaris, who speak Bakhtiyari Lurish, a dialect of Persian, are numerically the largest of these groups; they number approximately half a million, but only a third of the population remains nomadic. This still makes them the single most important seasonally migrant community in the country. Their homeland straddles the central Zagros Mountains of western Iran. The nomadic Bakhtiyaris winter in the foothills on the edge of the Khuzestan plain and spend the summer months on the opposite side of the Zagros Mountains in the lush pastures of the Chahal Mahal valley. The Bakhtiyaris are, therefore, a very significant example of traditional long-distance nomadism that continues to exist today in the Islamic world. The term Bakhtiyari derives from two Persian words which translate as 'companions of chance'; this might relate to the dangerous and difficult annual migrations these nomads make.

The complex social organisation of the Bakhtiyari tribe is crucial to its survival as a whole. The Bakhtiyaris were traditionally headed by an *il-khan*,⁴ whose assistant and subordinates would all play important roles in coordinating migrations and mediating disputes such as the overgrazing of pastures. To maintain some control over the tribes, Iranian governments for centuries would appoint and dismiss these tribal

2  According to Ibn Khaldun, '*asabiyyah* or 'group solidarity' exists at all levels of civilisation, but is stongest in nomadic societies. He maintains that each civilisation has within itself the seeds of its own destruction. When a nomadic community, that once lived on the periphery of a thriving state, absorbs that civilisation and becomes its new dynastic rulers, its level of '*asabiyyah* decreases with each new generation. This continues until the internal cohesion and bonds to the original nomadic group become so weak that the dynasty is no longer able to withstand outside forces wishing to repace it. This is how Ibn Khaldun explained the rise and fall of all civilisations. A classic example of Ibn Khaldun's '*asabiyyah* cycle was the Il-Khanid Dynasty (1256-1335) of Mongol rulers in medieval Iran, where nomadic invaders, within a generation, assimilated into and absorbed Islamic civilisation.

3  Towfiq, "'Ašayer (tribes)."

4  Each branch of the Bakhtiyari tribe was headed by a *khan*.

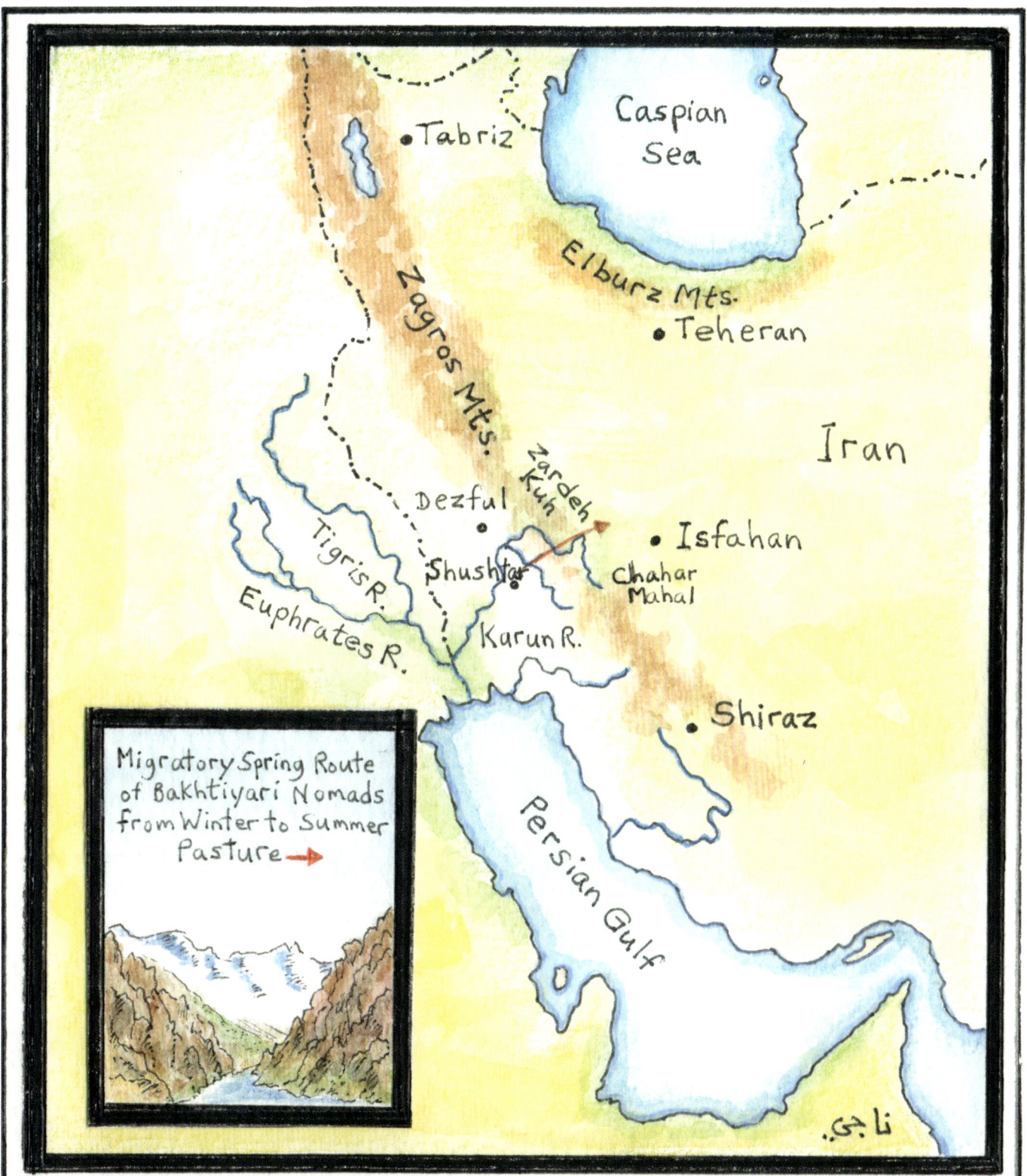

The horse is highly revered by all nomadic societies and is often the animal most beloved by them. On special occasions such as weddings, the Bakhtiyaris' horses are often decorated with beautifully woven horse blankets. Such coverings can also protect the horse from inclement weather.

heads. This became most pronounced during the Qajar (1785-1925) and later (1925-1979) Pahlavi periods of Iranian history.

The presence of nomads, like the Bakhtiyaris, in western Iran dates back to the time of Alexander the Great (fourth century BCE), or even before. Some historians believe the Mongol invasions of the thirteenth century CE forced the Bakhtiyaris and other nomadic peoples to inhabit the more remote regions of Iran.[5] By the time the Qajar rulers were replaced by Pahlavi ones, the reigning shahs had become suspicious of all nomadic peoples, especially the Bakhtiyaris. In 1923, Reza Shah Pahlavi was determined to restrict the freedoms once enjoyed by these nomads. He decided that their way of life had no place in his quest to modernise the country. Subsequently, he prevented the Bakhtiyaris from completing any seasonal migration and was thus temporarily successful in forcibly settling many nomads. Leading tribal khans were also removed and replaced by army officers, who were the Shah's 'yes men'. Not unlike the reforms made in neighbouring Türkiye by Mustafa Kemal Atatürk, Reza Shah outlawed the wearing of beards and traditional tribal clothing. He instead imposed the wearing of western-style clothing and head gear. The Land Reform Law of 1962 and the nationalisation of pasturelands (initiated by Muhammed Reza Pahlavi) obviously affected some Bakhtiyari tribal areas, where the land, as a direct result of these measures, was overgrazed. After the 1979 Islamic Revolution, Bakhtiyari tribal regions of Iran benefitted from the construction of roads, bridges, schools, houses and the installation of electricity lines to villages. Government incentives encouraged some nomads to volunteer to be resettled. But as of the writing of this book, tens of thousands of tribal members have not been persuaded to give up their independence as free-wandering nomads.

5 Khuzestan, the remote, arid province occupying the southwest corner of Iran was the Bakhtiyaris' traditional winter quarters. Interestingly, oil revenues (given by the Anglo-Persian Oil Company) from tribal land made the Bakhtiyaris a significant force in the history of Iran in the early twentieth century, until Iranian oilfields were nationalised in 1951. The presence of oil on Bakhtiyari territory motivated Reza Shah Pahlavi to undermine the autonomy of the nomads.

The economic mainstay of most tribes in Iran is sheep and goat breeding. Larger animals such as camels, horses and donkeys are also bred, essentially as pack animals. The most important animal products provided by nomads to the sedentary markets include wool, goat-hair, dried balls of sour yoghurt (*kashk*), and lamb meat.[6] Until the twentieth century, the Zagros Mountains and the surrounding environs were home to a wide range of wild beasts including lions, snow-leopards, boars, bears, hyenas and wolves. As seasonal migrants, the Bakhtiyaris became extremely familiar with their environment and all the fauna and flora found there. Today, they continue to be avid and proficient hunters of gazelles, ibexes and wild fowl, like pheasants. Wild plants, such as mushrooms and various berries provide an additional source of food as well as natural dyes for the wool used in tribal weavings. The creative, artistic and competent Bakhtiyari weavers prefer high quality wool and hair from sheep or goats to produce stunning works of tribal art that are not only practical, but also beautiful. Such items include hand-made rope, *malbands* (straps), saddle bags, storage bags, floor coverings and dining cloths (*sofras*). These trappings of the nomadic lifestyle are sometimes the only tangible connection people in the West have with the rich cultural heritage of Muslim nomads. As authentic migratory nomads, and all that their traditional lifestyle entails, become a rarer phenomenon throughout the Islamic world, these utilitarian hand-made works of art are being keenly sought after by collectors.[7]

The spring migration of the Bakhtiyari nomads from their winter quarters (*qishlaq* in Turkish, *garmsir* in Persian) on the

▲
The creative, artistic and competent Bakhtiyari weavers prefer high quality wool and hair from sheep or goats to produce stunning works of tribal art that are not only practical, but also beautiful. This is an example of a *kilim* woven to cover the floor of a Bakhtiyari tent.

6  The upheavals in the aftermath of the 1979 Iranian Revolution resulted in a severe shortage of fresh meat throughout the country; some nomads were thereby encouraged to return to their pastoral lifestyle to satisfy the demand for this essential commodity.

7  The author hopes that readers of the present work will appreciate the innate beauty of these tribal trappings produced by pre-agrarian, carbon-free societies that often inhabit some very inhospitable environments.

Bakhtiyaris of Western Iran 83

The traditional Bakhtiyari tent is made from black panels of woven goat hair, which retain heat and repel water during the winter. The interiors are sparse, but functional – filled with a colourful array of woven and embroidered decorative trappings like salt bags, saddle bags, storage bags, *malbands*, prayer rugs, horse blankets and saddles. In one corner food is cooked over a simple hearth and unleavened flat bread is baked; in another, a *kilm* is woven on a horizontal loom.

plains of Khuzestan to their summer quarters (*yaylaq* in Turkish, *sardsir* in Persian) in the green pastures of the Chahar Mahal valley of the Zagros Mountains has been carefully documented by both Muslim and non-Muslim researchers. The arduous 300 km migration (*kuch* in Persian) starts in late March and is completed by trekking from dawn until dusk each day for six to eight weeks. Migration is necessary because the winter quarters become unbearable during the summer months with all the grassland parched; similarly, the summer highlands become uninhabitable during the winter months from October to April. Migration thus appears to be the only solution for those nomads who wish to raise sheep and goats in this remote part of Iran. Following well-trodden paths across mountain gorges and peaks, and traversing the fast-flowing icy waters of Karun River with tens of thousands of animals is no mean feat. As there is no part of western Iran that has ample grassland year round, the Bakhtiyaris and other tribes will graze their animals on the western side of the Zagros Mountains in winter, and will migrate to the alpine pastures on the eastern side of the mountains in summer.[8]

As can be imagined, the Bakhtiyari nomads must travel light: heavy tents and other trappings are left behind in the winter quarters as tents previously brought to the summer quarters await them.[9] Only the absolute necessities are transported along with hundreds of thousands of animals (chickens, goats, sheep, donkeys, horses and mules).

The yearly migration of the Bakhtiyari nomads is truly a triumph of endurance. The most grueling and dangerous of all the obstacles to overcome must certainly be the fording of the Karun River. In late spring the river is wide and swollen with the melted

8  Neighbouring Luri and Kurdish tribes will often migrate with their herds along routes used by the Bakhtiyaris.

9  The traditional Bakhtiyari tent is made from black panels of woven goat hair, which retain heat and repel water during the winter. The interiors are sparse, but functional – filled with a colourful array of woven and embroidered, decorative trappings like salt bags, saddle bags, storage bags, *malband*s, prayer rugs, horse blankets and saddles. In one corner food is cooked over a simple hearth and unleavened flat bread is baked; in another, a *kilm* is woven on a horizontal loom.

Bakhtiyaris of Western Iran 85

The spring migration of the Bakhtiyari nomads from their winter quarters on the plains of Khuzestan to their summer quarters in the green pastures of the Chahar Mahal valley of the Zagros Mountains is a grueling trek. Even in late spring, the Zagros peaks are snow-capped, and thousands of nomads with their tens of thousands of animals must traverse them before they can reach their goal: the lush mountain pastures.

snow from the lofty Zagros peaks. But the seemingly impossible task is accomplished year in and year out. In 1925, an American author and film maker, Merian Cooper, experienced for himself the hardships of the Bakhtiyari migration and faithfully recorded this feat of endurance on film (*Grass: A Nation's Battle for Life*) and in the book *Grass*.[10]

> Here's a river a half-mile wide. Its waters are swelled to a rushing torrent by the melting snows of a hundred mountain peaks. The river is icy cold. It is filled with whirlpools, cross-currents, rapids. It is tearing through mountain gorges with cliff-like jagged shores, and it is bridgeless and boatless. Here's the problem. On one side of the river are five thousand people with all their worldly goods and perhaps fifty thousand animals. There are women here, children, babies. It is spring and the herds and flocks and any number of baby animals. The people have no boats. But they must cross and cross quickly. There's little or no grazing for the animals on this side of the river. They must cross!

> ... Every tribe has in its saddle-bags quantities of goatskins. A score of skins are blown up, tied and then fastened to one side of rows of long sticks. A heavy rug is laid on top. There's your raft.[11] There's your boat for the women and children and baby animals.[12]

The migration has to have some organisation if it is to be successful. It is, therefore, the duty of the tribal leader, the *il-khan*, to divide the tribe into groups (of approximately 5,000 nomads), each headed by a migration guide, or *il-rah*. He will preside over the river crossings (which can take five or six days) and treks into

10 Cooper and two companions successfully completed the entire length of the Bakhtiyari spring migration following in the steps of the nomads.

the mountains.

Once the rivers have been forded, there is still another hurdle to surmount before the Bakhtiyari nomads reach their ultimate goal: the alpine pastures of the Chahar Mahal valley on the eastern slopes of the Zagros Mountains. The snowy peaks of several Zagros Mountain ranges must now be crossed by zigzagging up and down their faces on foot. The *il-rahs* lead their fellow tribesmen along paths traversed by countless generations of Bakhtiyari nomads. Until recently, a group of men would act as a scouting party to 'break trail' in the deep snow with their bare feet in order to ease passage for the rest of the tribe and all the accompanying animals.[13]

> On descending the heights of the mountains, the nomads could see the green pastures of the Chahar Mahal valley below. "At last, snow and mountain ended… Here out to the horizon stretched green valleys through which, in the golden sunshine, rippled silver streams feeding the luxuriant young grass. Here was the prize of the gallant fight. Here was the land of plenty."[14]

Once safely camped in the green pastures of the *yaylaq*, the Bakhtiyaris can afford to briefly rest while their animals comfortably graze. In the evening, campfires are lit using animal droppings as fuel, and the nomads converse or listen to recitations of epic Persian poems. Simple, but savoury food is prepared: herbal omelettes cooked in sheep butter, thick herbal yoghurt, freshly baked thin sheets of unleavened bread and plates of sticky dates.

After spending four months in the mountain pastures, the

---

11  Today, these precarious goatskin rafts mentioned by Cooper are no longer used in crossing the Karun River. Some livestock are now transported by truck.

12  Cooper, *Grass*, 219-229.

13  Bashiri, "The Iranian Tribes."

14  Cooper, *Grass*, 358-359.

Finally, the goal of the Bakhtiyari nomads' grueling migration is reached: the green pastures of the Chahar Mahal Valley, the 'land of plenty'.

Bakhtiyaris migrate back to their winter home in Khuzestan by following the same route, through mountain gorges and across rivers. Despite the many incentives today for nomads to relinquish their traditional lifestyle, there are still many Bakhtiyaris willing to endure these hardships and continue living as their ancestors of yore – as 'people of the wind'.

# 8

# QASHQA'IS
## OF SOUTHERN IRAN

Turkish and Persian-speaking nomadic tribes have coexisted in Iran for centuries.[1] Today, the Qashqa'is form the largest Turkish-speaking tribal confederacy in Iran.[2] They, along with the Bakhtiyaris, are the nation's two largest groups of migratory nomads.

1. The Qashqa'is (who number approximately one million) speak an ancient dialect of Turkish, related to those spoken by the Turkmens of north-eastern Iran and the Azeris of north-western Iran. Qashqa'is also speak *Farsi* (Persian), the national language of the country.

2. Although the Qashqa'is speak Turkish, the confederacy is comprised of approximately two hundred tribes of diverse ethnic origins: Turkic, Arab, Kurdish and Luri.

The Qashqa'i tribes, like all Turkic-speaking tribes, migrated west from the Altai Mountains of Central Asia, the ancestral home of all Turkic peoples. When this migration actually occurred is as controversial as the origin of their name. Some believe it derives from the Turkish verb *kaçmak* ('to flee'), or possibly from the name Kashgar, the oasis city and major trading emporium in East Turkistan. What is more likely, however, is that the Qashqa'is migrated to Iranian territory in the wake of the Mongol conquest of western Asia in the thirteenth century.[3]

Qashqa'i tribal lands are today located in the Zagros Mountains and the neighbouring plains of south-western Iran. These nomads proudly claim the *Ak Koyunlu* ('White Sheep') tribe as their distant ancestors. The *Ak Koyunlu* nomads are renowned for being the only tribe that was able to defeat Timur (died 1405). A century later, Iran was united for the first time by Shah Ismail, the Azeri Turkish founder of the Safavid dynasty (1501-1736). According to Qashqa'i legend, Shah Ismail requested Qashqa'i tribesmen to resettle in southern Iran to defend the country from a possible invasion by the Portuguese, who had arrived in the Persian Gulf. For five centuries, the Qashqa'is have continued to inhabit this region of Iran. The Confederacy of Qashqa'i tribes likely dates to the time of the Safavid ruler, Shah 'Abbas I (1587-1619), who granted the Qashqa'i control over all the tribes of Fars Province.

Throughout their modern history, Qashqa'i nomads have endured the same suffering as other nomadic groups as a result of government policies. In the 1930s, Reza Shah Pahlavi outlawed tribal nomadism throughout Iran; traditional tribal dress was banned as was the erection of tents and yurts. The single pass

3 Some Qashqa'is claim descent from a tribe transplanted by Hulagu Khan (grandson of Genghis Khan), who in 1256 devastated parts of Central Asia, including Kashgar and parts of Afghanistan. Yet another theory has the Qashqa'i tribe inhabiting Eastern Turkistan (Xinjiang, Western China) and leaving the area between the eleventh and fourteenth centuries. Timur supposedly transported some *Ak Koyunlu* tribesmen from Anatolia (eastern Türkiye) to eastern Iran. A group of these nomads fled (*qashqa'i*, meaning 'those who fled') south to Fars Province, where they survive until today.

connecting the nomads' summer and winter quarters was forcibly closed, thus curtailing the semi-annual migrations (*kuch*). This resulted in the deaths of many nomads and their animals.

The abdication in 1941 of Reza Shah in favour of his son was welcomed by many nomads who immediately resumed their traditional way of life. In 1956, the Qashqa'i Confederacy was disbanded, and by 1962, Muhammad Reza Pahlavi, once in full control of the government, began to implement revolutionary land reform policies that ultimately denied the Qashqa'is grazing rights to important pastures. Qashqa'i nomads rejected these land reforms and were further isolated when their two most powerful tribal chiefs, Nasir and Khosrow Khan, were exiled to America and Europe, where they remained until the 1979 Islamic Revolution.[4] They returned to Iran, but realising that they could not influence the revolution, the two brothers led an unsuccessful uprising against the new Khomeini government. Nasir Khan fled the country with the help of supporters, but his brother, Khosrow, was captured by the Revolutionary Guards and publically executed in Shiraz.[5]

Like their Bakhtiyari nomadic brothers, the Qashqa'is have twice yearly traversed portions of the Zagros Mountains of western Iran for centuries. Typically, ten to twelve families migrate together along with their flocks of goats and sheep. Each group of nomads is guided by an especially appointed khan for the duration of the two to three months needed to migrate between the winter and summer quarters. The tribal chief also assigns them a migration route and a special section of pastureland.[6]

---

4  Mohammad Reza, the last Pahlavi shah, unlike his father, never directly ordered the murder of any tribal chief. He still harboured a deep distrust, however, of all nomads whom he considered an impediment to his plans for making Iran a modern, oil-rich state. Muhammad Reza was particularly suspicious of the Qashqa'is after they had supported Mossadeq, (the popularly elected prime minister, who was ousted from office by a CIA-led coup) because of his challenge to the Shah. It is interesting that in 1962, Muhammad Reza Pahlavi's government officially declared that tribes did not exist in Iran.

5  Nasir Khan died in self-imposed exile in America in 1984, thus truly ending the influence of Qashqa'i khans in the Iranian government.

6  All nomadic migrations must be carefully organised, almost like a military campaign, if they are to be successful.

The spring migration of Qashqa'i nomads is a very colourful sight. Horses, camels and donkeys are used for transportation, and many are festooned with decorative saddle cloths and collars. The nomads' worldly belongings are heaped on their animals with tent poles projecting like knitting needles from a ball of wool. Qashqa'i women are known for their beautiful tribal dress – long, flowing floral-patterned skirts and veils that also offer protection against the incessant dusty winds.

The migration of Qashqa'i nomads is a very colourful sight. They begin their trek in early spring from their warm winter quarters (*qishlaq*) on the lowlands,[7] to the north of the Persian Gulf, and travel 500 km north to their summer quarters (*yaylaq*), the lush Zagros meadows, 4,000 metres above sea level. Horses, camels and donkeys are used for transportation and many are festooned with decorative saddle cloths and collars. The nomads' worldly belongings are heaped on their animals with tent poles projecting like knitting needles from a ball of wool. Qashqa'i women are known for their beautiful tribal dress – long, flowing floral-patterned dresses and veils that also offer protection against the incessant dusty winds.

An evocative account of Qashqa'i nomads on the move is given by the late British traveller, Bruce Chatwin:

> The Qashqa'i men were lean, hard-mouthed, weather beaten and wore cylindrical hats of white felt. The women were all in their finery: bright calico dresses bought especially for the spring journey. Some rode horses and donkeys; some were on camels, along with the tents and tent poles. Their bodies ebbed and flowed to pitching saddles. Their eyes were blinkered to the road ahead.
>
> A woman in saffron and green rode on a black horse. Behind her, bundled up together on a saddle, a child was playing with a motherless lamb; copper pots were clanking and there was a rooster tied on with a string... Her breasts were festooned with necklaces of gold coins and amulets. Like most nomad women, she wore her wealth.[8]

---

7   As the spring migration often coincides with the vernal equinox, the traditional beginning of spring in the northern hemisphere, there is often a festive mood amongst the nomads. This time of year is celebrated as *Nowruz* ('New Year'), the ancient Iranian new year festival, beginning on March 21.

8   Ure, *In Search of Nomads*, 62.

The migratory Qashqa'is prefer routes off the beaten track – far from inhabited regions and roadways.[9] They can often be seen passing by the ancient ruins of Persepolis, sixty kilometres outside of the city of Shiraz.[10] The Qashqa'is, like most nomads, are skilled horsemen and take great pride in breeding quality stallions. The horse is so essential to the nomads' prosperity that the Qashqa'is often say that "a good horse is like a member of the family."

Although annual migrations continue, newer obstacles – asphalted roads, extensive agribusinesses and irrigation projects – all hinder the once free passage of nomads along ancestral routes.[11]

The Qashqa'is are renowned for their exceptionally fine tribal weavings. The expertise of a nomad's weaving skill is a good gauge for determining the general civilisation of a nomadic people.[12] The high quality of these products is attributed to the wool, which is harvested from sheep that have grazed in the mountains and valleys near Shiraz.[13] The nomads' tents are bedecked with a glorious array of superb handmade weavings: saddle bags, salt bags (*namakdan*), tool bags, utility bags, dowry chest covers, *malbands* (wide woven straps) and tent pole covers.[14] Some weavers continue to only use wool coloured with vegetable (or even insect) dyes. The size of tribal weavings are restricted only by the size of the horizontal looms, which for practical purposes, cannot be as large as the vertical ones used by sedentary weavers.

During the 1960s and 1970s, the Qashqa'i traditional way of life was being forcibly affected by an increasingly intolerant government, who believed strongly in "sedentarisation through education". To encourage tribal elders to consider sedentary

9 With millions of sheep and goats on the march, Qashqa'is intentionally avoid populated areas where their animals may cause damage.

10 Persepolis was the ceremonial capital of the Persian Achaemenid Empire (ca 550-330 BCE). Today, the city's ruins are an impressive sight on the dusty, arid plain. Historians believe in 330 BCE Persepolis was looted then burned down by Alexander the Great as an act of revenge for the burning of the Acropolis of Athens by the Persians a century and a half before.

11 Beck, "The Qashqa'i of Iran."

12 Ure, *In Search of Nomads*, 57.

13 In the past, carpets were made from wool only; this necessitated the use of three kilos of washed and spun wool for every m² of carpet. (Towfiq, F. "'Ašayer (tribes).")

14 *Malbands* can be up to ten to twelve metres in length. Despite their decorative features, they have a very important practical use: they literally help support the wall of the circular *yurt* from collapsing. When migrating, nomads may use these same *malbands* to secure loads on pack animals' backs. The finely-woven *malband* would not be as abrasive as a rope and would thus spare the animal the pain of rope burn.

The migratory Qashqa'is prefer routes off the beaten track – far from inhabited regions and roadways. They can often be seen passing by the ancient ruins of Persepolis, sixty kilometres outside of the city of Shiraz. Sometimes, the spring migration coincides with the New Year (*Nowruz*), the ancient Iranian new year festival, beginning on March 21.

residence, the Shah established tribal dormitory schools for their children in urban centres, such as Shiraz.[15] Qashqa'i nomad children were permitted to wear their traditional dress at these segregated schools and were able to visit their parents during vacations. But these schools aimed to assimilate tribal children into the modern Iranian society; respect and obedience to country and its benevolent monarch were, therefore, instilled in pupils from the first grade. Instruction was in *Farsi* (Persian) only and the skills taught intentionally deprived them of ever becoming comfortable with the lifestyle of their parents and grandparents.

A contemporary Iranian scholar, Soheila Shahshahani, contends that the tribal-school/tent-school project pursued an "oppressive pedagogy" that only served the "dominating class in the society". In interviewing university students who had gone to tribal schools, this same researcher recounts that they felt ill-equipped as *Farsi* was not their mother tongue. Clearly, the first aim of the project was to make urban dwellers of tribal children and then to humiliate anyone who still had any fond attachments to the nomadic lifestyle.[16]

It was initiated throughout Iran in 1954, and directed by the American-educated Qashqa'i tribesman, Mohammad Bahmanbegi – the ideal person to lead the project, which was supported by the American government. Children up to the age of twelve were required to attend schools which were erected wherever the nomads' encampment might be. The school teachers were from the same ethnic group as the children, and both tent-school and teacher would migrate together along with the nomad families. The white canvas tentschools – distinct from the ubiquitous black goat-hair ones of the nomads – were pervasive throughout rural

15 While resident in Shiraz in 1973, the author observed the Shah, Muhammad Reza Pahlavi, inspecting the Qashqa'i Tribal School there. Some graduates of this school went on to receive scholarships for further education at Iranian universities. Today, some Qashqa'is proudly laud their tribesmen who eventually received graduate degrees – some in medicine. Many others, however, see the tent-school project as a deceptive and manipulative scheme to totally isolate tribal children from the traditional nomadic lifestyle of their ancestors.

16 Shahshahani, "Tribal schools of Iran: Sedentarisation through education.", 145.

areas of Iran, but not all provided the same level of education. Unfortunately, some tribal groups, such as the Baluchis, have suffered from absentee teachers and insufficient school supplies.

Today, in the twenty-first century, nomadism is increasingly viewed by governments in the Islamic lands as a curiosity – an exotic relic of the past. Nomads are sometimes even paraded in their finery for the benefit of inquisitive tourists. But around campfires in some nomadic communities, epic tales of tribal valour are still being told. And some youth continue to learn valuable survival skills, handed down over the centuries from father to son. Nomadism is inherently protective of the environment. The Qashqa'is, for example, live in a fragile one that is being radically affected by human interference from outside tribal territory. Over hundreds of years, they have fine-tuned their knowledge of ways to deal with droughts and flooding – today, often being the direct result of climate change. They try to maintain a sustainable lifestyle by protecting their environment from overgrazing and the overhunting of local wildlife.

Centuries ago, Ibn Khaldun identified a symbiotic relationship between pastoral nomadic tribes and sedentary society. Even though nomads lived on the margins of urban culture, they contributed in some part to its economy. Even today, in post-revolutionary Iran, some Qashqa'is are profiting economically due to the country's insatiable demand for fresh meat and dairy products. As sedentary society continues to embrace the

▲

The Qashqa' is are renowned for their exceptionally fine tribal weavings. The nomads' tents are bedecked with a glorious array of superb handmade weavings, such as salt bags (*namakdan*). In the more arid regions of the Middle East and Central Asia, animal survival depended on rock salt; consequently, these beautifully woven bags were used to store this precious commodity. During long migrations, shepherds carry salt for their animals in these bags. The narrow bottle-neck can be folded to prevent lumps of salt from spilling out.

consumption of organic produce such as free-range poultry, eggs, grass-fed meat and free-range dairy products, one hopes there will always be a market for the natural produce offered by pastoral nomads.

    The world's population is expected to reach ten billion by the year 2050. Urban congestion and air pollution worldwide are making many cities uninhabitable and concerningly unsustainable. Mega-cities are now becoming hyper-cities, which in turn are morphing into unimaginable super-cities.[17] Although sustainability has become the century's byword, many people would consider their world to be entirely unsustainable without 24/7 access to mobile phones and social media – their lifeblood. We can all profit from appraising the humble, sustainable lifestyle led by the Qashqa'is and other Muslim nomads of the Islamic world.

---

17  A mega-city, like Los Angeles (19,000,000), has a population between 10-20 million; a hyper-city, like Greater Jakarta (30,000,000), has a population between 20-40 million; a super-city, like the Pearl River Delta, China (49,000,000), has a population exceeding 40 million.

In an attempt to sedentarise nomads through modern education, the Shah of Iran initiated the tribal-school/tent-school project in 1954. All children up to the age of twelve were required to attend schools which were erected wherever the nomads' encampment might be. The school teachers were from the same ethnic group as the children and both tent-school and teacher would migrate together along with the nomad families. The white canvas tent-schools – distinct from the ubiquitous black-goat hair ones of the nomads – were pervasive throughout rural areas of Iran.

# 9

# KUCHIS
## OF SOUTHERN AFGHANISTAN

Like neighbouring Iran, Afghanistan has for centuries been home to bands of pastoral nomads, but because of the recent decades of tumultuous political unrest, they have suffered differently from their nomadic brethren in bordering countries. The 1978 Communist coup d'état, the subsequent invasion and occupation of the country by the world's two superpowers and the ensuing upheavals caused by the civil war, have all had devastating consequences for the indigenous nomadic populations of Afghanistan.

Landmines placed along migratory routes and pasturelands, for example, have seriously affected their ability to continue their traditional lifestyle. The Persian word *kuchi* is a generic term applied to Afghan nomads.[1] Although there are many ethnolinguistic groups in Afghanistan who still practice mobile pastoralism, the Pashto-speaking Kuchis are the most populous.[2] Due to the ravages of war, no precise population figures exist, but there are estimated to be approximately 1.5 million Kuchis who remain fully nomadic.

Interestingly, few nomads actually call themselves Kuchi; the term has been used by foreigners to identify the nomadic population of Afghanistan, who live in black goat-hair tents.[3] A Kuchi might refer to himself as either a *watani maldar* (local herd-owner) or a *kuchi maldar* (migrating herd-owner) – the true nomad.

Not unlike Iran, whose nomadic peoples have spawned several dynasties of rulers, many Afghans also have nomadic roots. "Long before there was a nation state called Afghanistan, there were Kuchis, traveling a vast land that is hospitable only part of the time…

Most Afghans of Pashtun heritage and many Afghans from ethnic minorities can trace their lineage to nomadic ancestors."[4] For example, Ashraf Ghani, the president of Afghanistan (elected in September, 2014) is ethnically a Pashtun hailing from the Ahmadzai tribe, which has many members who live as Kuchis, migrating throughout the provinces of southern Afghanistan. Most Kuchis are mobile pastoralists, who move with their fully loaded pack camels and flocks of goats and sheep, but some are semi-sedentary and a few are nomadic traders. While smaller nomadic groups inhabit the northern parts of Afghanistan,

---

1  *Kuchi* is derived from the Persian word *kuch* ('migration'; *kuch kardan* 'to migrate'); *kuchi*, therefore means 'one who migrates'. The term has passed into Turkic languages, like modern Turkish, as *göç* ('migration'; *göç etmek*, 'to migrate').

2  Other nomadic peoples besides the Pashtun Kuchis include the Turkic Aimaqs, Baluchis, Kyrgyz, Turkmen, Uzbeks and the semi-nomadic Pamiri peoples of Afghanistan's remote north-eastern Wakhan Corridor.

3  The term Kuchi has been used as a pejorative (resembling the use of the word 'gypsy' in the West) by many urban Afghans and others who are either embarrassed by this element in their 'modern' society, or hold grievances against them.

4  Baldauf. "Nomads in a Troubled Land."

Afghanistan has for centuries been home to bands of pastoral nomads who migrate across the sweeping varied landscapes – mountains, deserts and steppes – of this ancient land-locked country. In spring and autumn, the infinite sight of their caravans of camels or yaks slowing, crossing and dotting this rugged terrain is remarkable.

In the past, a fine balance was maintained between nomads and the environment, which enabled them to continually move to newer pastures – leaving the delicate ecosystem unharmed. Some Kuchis were also important traders, linking Afghanistan with the outside world. They traditionally migrate in small units of about ten households along with 500–600 goats and sheep. In the past, nomads would often cross borders with impunity. However, today these seasonal migrations have been made more precarious because of the ongoing political unrest in the region.

the Kuchis traditionally traverse the mountains and valleys of the Pashto-speaking southern regions of Ghazni, Kandahar, Helmand and Nimroz.

In 2011, an Italian academic had an engaging conversation with a local Kabul resident, who unreservedly divulged his dislike of nomads. "Kuchis would claim that they were the real Afghans, as the first inhabitants of Afghanistan were nomadic in ancient times. Thus it is to be considered that all of Afghanistan once belonged to the Kuchis: Indeed, they would have it that actually the whole world belongs to them, as all of humanity was once made of nomads!"[5] While this urban Afghan is obviously exaggerating, the fact remains that nomads have inhabited this part of Asia for hundreds, if not thousands of years. The precise origin of the contemporary Kuchis remains a mystery; some claim they were indigenous to the northern regions of the country, and were forced to retreat south when the Mongols first invaded this part of the world in the thirteenth century.

For the *kuchi maldars* (the mobile pastoralists), the centuries-old annual spring migration across Afghanistan's fragile terrain – from the arid lowlands to isolated mountain pastures – was the means for their survival. The unimpeded grazing of their herds would enable them to produce the dairy products, fresh meat and wool that could be sold or bartered for in the nearest village or town. Kuchis traditionally migrate in small units of about ten households along with 500-600 goats and sheep. In the past, nomads would often cross borders with impunity, and during the course of the migration, nomads from other tribes would often travel together and even share grazing areas.[6] However, today these seasonal migrations have been made more precarious because of the ongoing political unrest. Nomadism in its essence

[5] Foschini, "The Social Wandering of the Afghan Kuchi."

[6] Ker, "Singing in the Wilderness: Kuchi Nomads in Modern Afghanistan."

requires free, unrestricted access to pastureland. Afghanistan is a rugged mountainous land with inadequate water resources to sustain herds of grazing animals for any length of time. This forces its nomads never to dwell in one area for long.

The British explorer and travel writer, Freya Stark, wrote evocatively of the migratory Kuchi women she saw in the late 1960s in preconflict Afghanistan. At that time, the Kuchis were prevalent – their winding caravans of pack camels,[7] herds of goats and sheep dotted the Afghan landscape in spring and autumn.

> Some held a child in one arm, while they stretched the other to manage the camel's neck with a long wand of poplar or willow; some had their shifting platform widened with brass and copper cauldrons, mattresses and quilts, beaked coffee pots, or gourds and skins for water, and livestock, new-born lamb, or cats and fowls tied by the leg in any receptacle that had come to hand.[8]

In the past, a fine balance was maintained between nomads and the environment, which enabled them to continually move to newer pastures – leaving the delicate ecosystem unharmed. As the amount of grazing land shrinks, the sustainability of these seasonal migrations is now in serious doubt. Severe droughts and environmental degradation have reduced the amount of practical grazing land for Kuchis, and have directly contributed to the depletion of livestock.

The sedentary population of Afghanistan has always relied to some degree on nomads to supply them with essential products, but decades of warfare have interrupted the provision of these. For years, in more peaceful times, the Kuchis owned thirty per

[7] Camels can transport very heavy loads, but within limit. Experts describe how a camel "will 'grumble, growl and show resentment' and that no notice be taken of such behaviour; but the moment the camel becomes silent, it is an indication that the load has reached the maximum acceptable weight." (Ure, *In Search of Nomads*, 139.)

[8] Stark, *The Minaret of Djam*, 126.

Kuchis, like most nomads, produce a wonderful array of tribal weavings. This is a splendid example of a hand-crafted camel collar, which adorns the animal on special occasions like a wedding or the first day of a major migration.

cent of the country's goats, sheep and camels. They were also mainly responsible for the supply of slaughter animals, wool, ghee, and *quroot* (dried balls of sour yoghurt) to the national economy.⁹ Some Kuchis were also important traders, linking Afghanistan with the outside world. Pack animals would be loaded with items like tea, sugar and matches, which would be bartered for fresh vegetables and the like.

The Kuchis supplied the market with nutritious dairy products. In addition, cashmere from goats and fine quality wool from their sheep helped fuel the lucrative cottage industry of carpet making. While larger weavings such as room size rugs are made on vertical looms by village weavers, the Kuchis, using horizontal ones, produce a remarkable array of splendid hand-crafted utility items for their own use.

> Their arts must be portable, the same bright patterns with meaning repeated through the centuries, to be stitched on the neckbands of their camels, or woven in rugs and saddlebags and carpets for their tents; or to wear... on to the baby jacket of a son. These are the easy things to carry, and no one is much harrowed by choice. And who has ever seen an 'inferiority complex', or heard of a 'guilt complex' in a nomad? From these symptoms of captivity, traditions that have got mixed and entangled and lost their scale of value, the nomads are immune.¹⁰

The Kuchis are particularly vulnerable as their livelihood depends on being able to move freely across grazing lands and along age-old migratory routes that are no longer safe or, at times, accessible. Landmines planted in these areas (and in irrigation systems used by the Kuchis) during the decades-long

9  "Durable solutions for Kuchi IDPs in the South of Afghanistan: Options and Opportunities." *Asia Consultants International*, 15.

10  Stark, *The Minaret of Djam*, 126.

Kuchis of Southern Afghanistan 111

Although a Kuchi's outer apparel can be quite plain, women's dresses can be artistically decorated with extremely fine embroidery. And all members of the nomadic family, including infants, will wear beautifully embroidered skull caps, which exhibit exceptional weaving skills.

conflict that still stifles Afghanistan's development, pose a never-ending obstacle to all mobile pastoralists. This egregious form of warfare has decimated the nomads' livestock.[11] Landmines as 'hidden killers' have also killed and maimed many Kuchis; they indeed may be labeled the real weapons of mass destruction.

For centuries, there had been a relatively peaceful relationship between Kuchis and the settled populations of Afghanistan. However, more recently, serious tensions have arisen over the rights to grazing land. Despite possessing *firmans* (royal decrees) granting access to pastures, the Pashto-speaking Kuchis have come into conflict with the sedentary Persian-speaking Hazaras[12] in the central highlands. This ongoing Kuchi-Hazara dispute is in fact much more than a disagreement over land tenure - it conceals an uglier, more serious interethnic and sectarian divide.

These hindrances have made the Kuchis' nomadic way of life all the more unsustainable. Even the demand for traditional Kuchi produce has dwindled due to rapid urbanisation and increased dairy imports from Iran and Pakistan. Still, many Kuchis would prefer nomadism to settlement in squalid government-sponsored camps. The Afghan government is willing to help improve the lot of all mobile pastoralists in the country, but believes they must first be settled. Perhaps the government should be more realistic and begin to address some of the Kuchis' priorities before implementing any resettlement programme.[13] Fuller Kuchi representation in the Afghan government, along with its sincere willingness to preserve the distinct culture of the country's nomads, would be a good first step.[14]

Another more extreme view regarding the possible settlement of Kuchis is proposed by Louis Dupree, the imminent American

▲

Afghan nomads, like the Kuchis, have a fondness for the horse. Each nomad will have his own individual carefully-crafted horse whip – an heirloom that is passed down from father to son. This is one of the rare possessions that a nomad takes pride in.

11  According to one report from the 1990s, approximately 60 percent of the current flocks owned by a sample survey of 1,400 families were killed by landmines. (Shawn Roberts and Jody Williams. *After the Guns Fall Silent: the Enduring Legacy of Landmines*, p.13)

12  The Hazaras are the third largest ethnic group in Afghanistan. It is widely believed that they have Mongol ancestry, perhaps being directly descended from the army of Genghiz Khan. Although they speak a Persian dialect (*Hazaragi*), it does contain some distinct Mongolian words. The Hazaras are predominantly Shi'a Muslims.

13  For example, urgent issues such as tribal education and improving the economic and social conditions, and the general well-being of all nomads must be addressed. The promise of mobile schools and health clinics remains, for the most part, unfulfilled.

14  Ker and Locke. "Singing in the Wilderness: Kuchi Nomads in Modern Afghanistan."

archaeologist, anthropologist and scholar of Afghan culture. He believes a nomad is always a nomad and that nomadism, therefore, must be inherent in his DNA.

> Show me a nomad who wants to settle down, and I'll show you a man who is psychologically ill … Many Afghan officials believe that nomads genuinely desire to settle down if given the opportunity. Nomads, however, look on themselves as superior beings … Any nomad desiring to settle down would be considered a psychopath by his peers … The nomad continues to look on the farmer with contempt. Even after he becomes semi-nomadic, semi-sedentary and eventually fully sedentary, his pride of nomadic ancestry makes him feel superior to his agelong farmer neighbours.[15]

Unquestionably, the traditional nomads of Afghanistan have suffered during the decades-long instability in their mountainous homeland. Despite the billions of dollars in international relief aid that has been donated to the country in recent years, the Kuchis have remained on the last rung of the largesse ladder, receiving little to nothing in humanitarian assistance.

One hopes that if the war-ravaged country can achieve some stability, an honest dialogue might be had between state officials and representatives of the nomads, the most marginalised sector of Afghan society. Still, there will always be the die-hard nomads, who in response to offers of settlement by government bureaucrats will retort: "God created Kuchis to wander in deserts, valleys and mountains… and raise animals."[16]

15  Dupree, *Afghanistan*, 168, 179.
16  "Kuchi nomads seek a better deal." *Irin News*.

# 10

# KYRGYZ
## OF NORTHEASTERN AFGHANISTAN

Afghanistan is a multi-ethnic society hosting at least fourteen distinct ethnolinguistic groups.[1] Perhaps the most marginalised and least studied of those still leading a nomadic lifestyle are the Kyrgyz, who inhabit the mountainous north-eastern part of the country. As pastoralists, they seek grassland for their flocks of sheep, goats and yaks, as well as for their horses and Bactrian camels.

1   The various ethnic groups in Afghanistan include the Pastuns, Tajiks, Hazaras, Uzbeks, Turkmen, Aimaqs, Farsiwans, Arabs, Pashais, Nuristanis, Baluchis, Pamiris, Qizilbash, Hindus, Gujars and Sikhs.

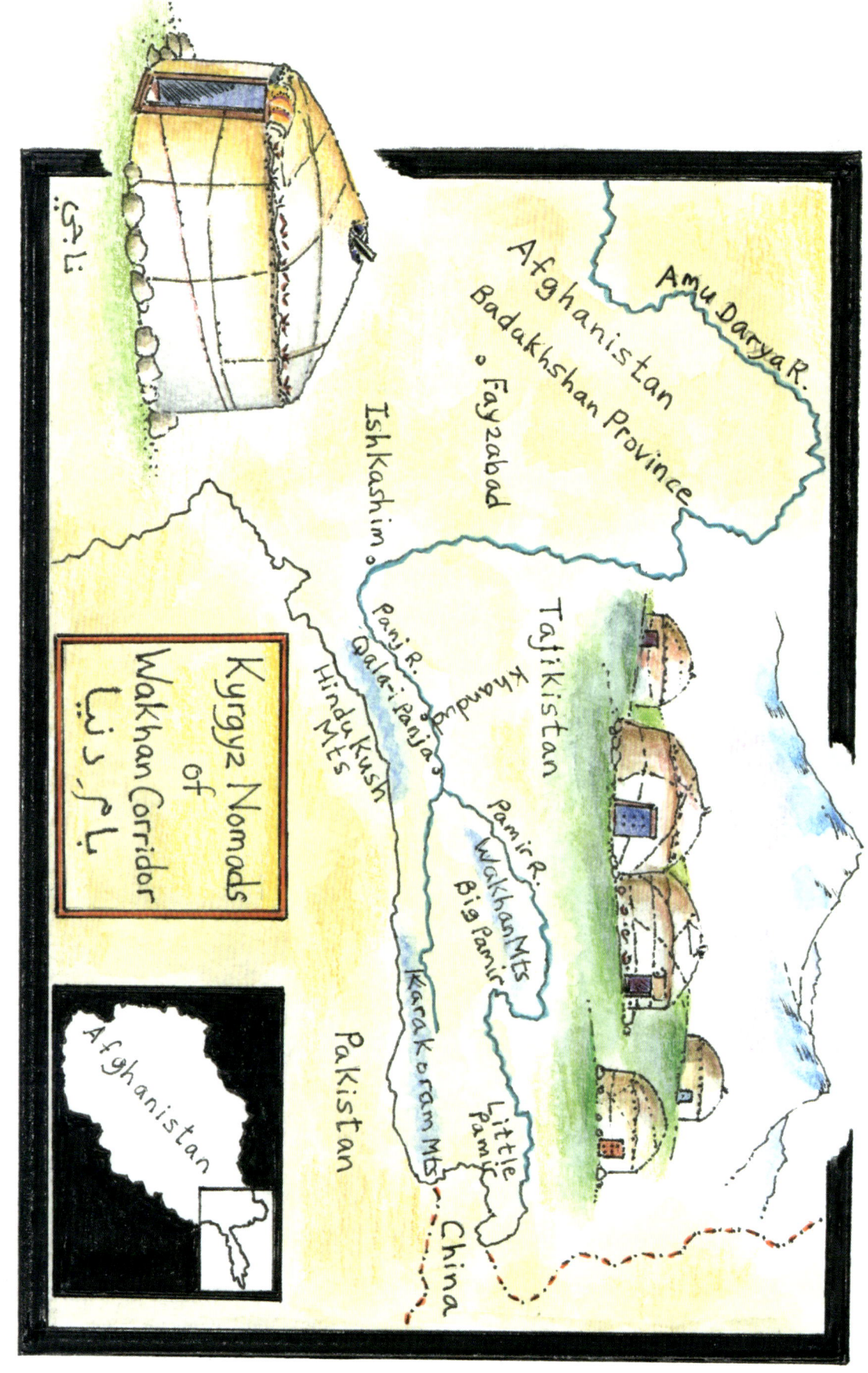

Life in the Wakhan Corridor of Badakhshan Province of Afghanistan is extremely harsh. At altitudes of 4,000 metres, the region has been named *bâm-i dunyâ*, or 'roof of the world', in Persian. The semi-nomadic indigenous Wakhi-speaking Isma'ilis, who live at lower elevations along the western section of the corridor, are the Kyrgyz nomads' only neighbours. Although the Kyrgyz are mostly self-sufficient, they do rely on traders and their semi-nomadic Wakhi neighbours for the few basics they lack, such as fresh vegetables, which the Wakhis grow in terraced fields irrigated from the Amu Darya River.

118  ENCOUNTERS WITH MUSLIM NOMADS

The author (in karakul cap) during his investigative visit to the Wakhan Corridor in August, 1974. In the background are the Pamir Mountains of Tajikistan. The Amu Darya River forms the natural border between Afghanistan and (Soviet) Tajikistan. The semi-nomadic, indigenous Wakhi-speaking Isma'ilis (seen in both photos), who inhabit the lowlands of the western section of the Wakhan Corridor, supply agricultural produce to the Kyrgyz nomads living in the high Pamir plateaus further east, close to the Chinese frontier. Because of its remoteness, this part of Afghanistan has been spared the turmoil of civil strife – resulting from foreign invasions and insurgencies – that for decades has plagued the rest of the country.

Life in the Wakhan Corridor of Badakhshan Province is extremely harsh.[2] At altitudes of 4,000 metres, the region has been named *bam-i dunya*, or 'roof of the world', in Persian. Here, three of the highest mountain ranges in the world meet in what is known as the 'Pamir Knot'.[3]

There are estimated to be between 1,500 to 2,000 Sunni Muslim Kyrgyz currently living in the eastern section of the Wakhan Corridor, the long and narrow panhandle that extends east to Afghanistan's only border with China. This rugged environment is one of the least hospitable inhabited regions on Earth. The semi-nomadic indigenous Wakhi-speaking Isma'ilis, who live at lower elevations (2,000-3,000 metres) along the western section of the corridor, are the nomads' only neighbours. At these high elevations, the mountain peaks are permanently glaciated. And as the Wakhan Corridor is beyond the reach of the monsoon rains that revitalise land to the south, it is devoid of abundant vegetation.[4]

Centuries ago, the Wakhan Corridor formed part of the Silk Road and long camel caravans loaded with the riches of the East would wind their way across its precarious peaks. For centuries, the Kyrgyz nomads roamed freely over the mountains and valleys of Kyrgyzstan, Tajikistan and western China, never confined to the small strip of land which is now their home. The high plateau meadows of the Wakhan Corridor are continuously fed by melting glaciers, thus making them covetable, rich grazing land. Nomads essentially controlled the steppes of Central Asia until the nineteenth century czarist colonial expansion and appropriation of their pastures. Russian farmers encouraged to settle in these areas, quickly ploughed under the grasslands that for generations had been the lifeblood of nomadic populations.

2   The Wakhan Corridor, the awkward-looking thin strip of territory jutting out from the north-eastern corner of Afghanistan proper, was a creation of the British and Russian Empires in the nineteenth century. At that time, they were each vying for economic and political influence in Central Asia. This century-long conflict for supremacy was known as the 'Great Game'. In 1895, the governments of the British Raj and Czarist Russia agreed that their respective empires should never share a frontier; thus, the panhandle was demarcated a neutral zone between them. The Panj River, the upper course of the important Amu Darya (Oxus) River, forms the natural border between Tajikistan and Afghanistan for 1,200 kilometres.

3   The three mountain ranges that converge to form the 'Pamir Knot' are the Hindu Kush (Afghanistan & Pakistan), the Karakoram (China, Pakistan & India) and the Pamirs (Afghanistan, Tajikistan, Kyrgyzstan, Pakistan & China).

4   For example, during the author's investigative visit to the Wakhan Corridor in August 1974, the very few trees he saw in the western 'lowlands' were hardy varieties of apricot, whose fruit was sun-dried and kernels crushed for their nutritional oil.

Perhaps the most marginalised and least studied of the nomadic populations of Afghanistan are the Kyrgyz, who inhabit the remote mountainous northeastern part of the country. As pastoralists, they seek grassland for their flocks of sheep, goats and yaks, as well as for their horses and Bactrian camels. There are estimated to be between 1,500 to 2,000 Sunni Muslim Kyrgyz currently living in the eastern section of the Wakhan Corridor, the long and narrow panhandle that extends east to Afghanistan's only border with China. This rugged environment is one of the least hospitable inhabited regions on Earth.

The recent history of Afghanistan's nomadic Kyrgyz population is one of perpetual movement in search of a sanctuary. Life became even more intolerable for them after the 1917 Bolshevik revolution. Within a few years of the founding of the Soviet Communist state, a new radical government policy was implemented "to eradicate for all time the political, economic and cultural backwardness of the nomad[ic] peoples and help them reach the level of the more highly developed peoples of Russia."[5] Between the years 1924-28, nomadism in Central Asia was brutally eradicated by the forcible internment of millions of nomads on land and in collective farms.

Of all Central Asian tribes, both Kyrgyz and Kazakh nomads were considered the most primitive – and consequently, the most in need of 'civilising' Soviet indoctrination. As a direct result of this targeting, the Kyrgyz sought asylum in the lofty plateaus of the Wakhan Corridor, hoping that this remote, unwelcoming habitat would be a safe haven, where they could preserve their traditional way of life. The geographical distance between the Pamirs and the Afghan government in Kabul was enough to keep them safe until 1946, when the Soviets conducted military raids across the northern border. The Kyrgyz were forced to flee anew, this time to the near-by Pamir region of western China. But by 1949, the Communist revolution there drove the nomads back to the isolation of Afghanistan's Wakhan Corridor. By that time, they were the only Kyrgyz living outside the clutches of Soviet and Chinese Communist control. They were once again in relative safety and free to pursue pastoral nomadism uninterruptedly. However, they were also totally isolated from their Turkic brethren in Soviet Central Asia and western China (Xinjiang).

5   Shahrani, "Afghanistan's Kirghiz in Türkiye."

The Kyrgyz yurt can be easily erected and disassembled. The wooden lattice frame, door, poles, *tunduk* and felt roofing can all be loaded on to a pack camel for transportation to a new location.

Having fled the Communist regimes of Soviet Russia and China, one can imagine the fear the Afghan Kyrgyz felt when the coup in Kabul in April 1978 installed a pro-Soviet Communist government. Within a year, the desire to preserve their Muslim and ethnic identities forced the Kyrgyz nomads to flee south to Gilgit, in northern Pakistan. Unfortunately, the warmer climate there negatively affected both the nomads and their herds of goats and sheep. The Kyrgyz nomads' khan, Haji Rahman Qul, next approached the Americans for permission for one thousand of his people to be settled in the more climate-friendly Alaska. But his request for visas was denied. By 1983, the government of Türkiye willingly accepted to resettle a total of three thousand refugees of Turkic origin, including most of the Kyrgyz nomads from Afghanistan. It must be said that not all of these nomads followed their khan to Pakistan, and of the hundreds who followed him into exile, not all travelled with him to Türkiye. Those who resisted vacating their Pamir habitat, continue their pastoral existence there despite all the ongoing chaos blighting the rest of Afghanistan.

The Kyrgyz nomads who still inhabit the high plateaus (Big [*Ulu*] Pamir & Little Pamir) of the eastern Wakhan Corridor are mostly self-sufficient and rely on traders and their semi-nomadic Wakhi neighbours in the western lowlands for the few basics they lack. As pastoralists, their lives are centred on tending sheep, goats, yaks and Bactrian camels, from which they obtain all their dairy products: milk, butter, yoghurt, and *quroot* (dried balls of sour yoghurt).[6] Meat is eaten only on festive occasions. By the careful management of their herds and grazing lands, the Kyrgyz have been able to support tens of thousands of animals. A complex network of trading has enabled them to sell some of their livestock even in far-off Kabul.

6  The thin soil of the western lowlands of the Wakhan Corridor produces some wheat, barley and vegetables – all grown by the semi-nomadic Wakhis. A unique type of wheat, seemingly impervious to the extreme cold, matures in only forty days. The Kyrgyz nomads' protein-rich diet is based on dairy products. In the long winters, a mixture of tea, salt, yak butter and milk is served boiling hot, along with a heavy unleavened flat bread made from wheat and pea/bean flour.

The Kyrgyz of the Wakhan Corridor are 'vertical' nomads, i.e., they migrate from low-lying winter quarters to higher summer ones. The high altitudes (exceeding 4,000 metres) of their habitat demand stamina from both man and beast during the annual long-distance migratory cycle. Although the horse is revered by all Kyrgyz, who proclaim they were born and raised on horses, yaks are ridden at these higher elevations.

▲

The wooden components of the yurt consist firstly of the frame, made from seasoned willow, birch or poplar branches. The interior of the yurt is a wonder to behold: hundreds of metres of magnificently woven *malbands* that are used to reinforce the wooden frame.

Until the Soviet invasion of Afghanistan in 1979, a very practical economic system (devised by their khan, Haji Rahman Qul) existed among the Kyrgyz. According to this *amanet* system, nomads could 'borrow' some of Rahman Qul's sheep, for example, and use their milk, wool and dung (for fuel). Any animals born became the property of the khan, however. At the end of each year, the nomads would have to replenish the khan's flocks by contributing a quota of lambs. Surplus animals were then shipped off to Kabul for sale with the proceeds being used to purchase provisions, such as rice, sugar and tea.[7] The Kyrgyz know their environment is a harsh one, but thank their Creator for the luxuriant pastures and fresh water of the Big and Little Pamirs that enable their herds to naturally increase in size.

The Kyrgyz of the Wakhan Corridor are 'vertical' nomads, i.e., they migrate from low-lying winter quarters (*qishlaq*) to higher summer ones (*yaylaq*). The high altitudes (exceeding 4,000 metres) of both the Big (*Ulu*) and Little Pamir regions demand stamina from both man and beast during the annual long-distance migratory cycle. Although the horse is revered by all Kyrgyz, who proclaim they were born and raised on horses, yaks are ridden at these higher elevations. Yaks, often touted as the 'ships of the snow', are remarkable in their adaptation to these high altitudes.[8] Because their larger lungs can absorb more oxygen than other animals, they are invaluable in transporting the nomads' belongings during the seasonal migration.

7  Bashiri, "The Kirghiz of Afghanistan."
8  Bashiri, "The Kirghiz of Afghanistan."

Most Central Asian nomads, including the Kyrgyz, live in collapsible circular dwellings called yurts. This dwelling has been used for centuries – perhaps millennia – with its design essentially unchanged. The roof is covered with felt matting and the whole wooden frame structure is reinforced with wide, beautifully woven malbands (tent straps). Balls of quroot (dried yoghurt, cheese or sour milk) are drying outside this yurt.

# Kyrgyz of Northeastern Afghanistan

▲
The most vulnerable part of the yurt is the door. Both door and frame, therefore, are usually made of very strong oak. A thick, beautifully designed felt roll-up cover provides excellent insulation from extreme cold and heat.

Most Central Asian nomads live in collapsible circular[9] dwellings called yurts.[10] This dwelling has been used for centuries – perhaps millennia – with its design essentially unchanged. It is an extremely practical structure that can be easily dismantled, loaded onto pack animals and re-erected at a new camp ground. It is also sustainable as only natural readily available materials are used in its construction: wood for the frame and felt (pressed wool) for coverings. No nails are used as all parts are perfectly crafted to fit together. These materials offer excellent insulation for making the yurt cool in summer and warm in winter. The wooden components consist firstly of the frame, made from seasoned willow, birch or poplar branches. The *tunduk*, the all-important 'roof ring', is also made of wood, as are the door and threshold.[11]

When nomads arrive at a destination, their yurts can be quickly assembled. First, the wooden door is placed facing east to catch the first rays of the sun. The door lintel is intentionally low, which compels the visitor who enters to bow his head in respect to the owners. Next, the *tunduk* is erected on the top of a long pole. This provides the opening at the top of the yurt from which smoke can escape and light and fresh air can enter. The *tunduk* is the one part of the yurt that does not need to be frequently replaced. It is, therefore, often handed down from father to son as an important family heirloom. The bent wooden branches are then positioned to all meet at the *tunduk*, roof ring. Finally, once the framework is in place, it can be covered with a felt 'skin', which is lashed to the wooden poles with sturdy, hand-woven straps, or *malbands*. In bad weather, a special square piece of durable felt matting is placed

---

9   It has been suggested by some historians that the domed Central Asian yurt might have influenced early Islamic architecture.

10  The term 'yurt' derives from the Turkic word meaning 'dwelling place' or 'homeland'.

11  The word *tunduk* means 'window' and actually acts as a window – an opening to the clear blue sky and sunshine.

When nomads arrive at a destination, their yurts can be quickly assembled. After the wooden door has been erected, the roof ring, or *tunduk* is inserted on the top of a long pole. This provides the opening at the top of the yurt from which smoke can escape and light and fresh air can enter. The bent wooden branches of the frame are then positioned to all meet at the *tunduk*. The most fearful curse of Kyrgyz nomads is *"tushsun"* – "may your yurt's *tunduk* fall (collapse)"!

over the *tunduk*. The number of layers of felt covering depends on the season, but the outer sheeting is always water-proofed with oil.

The interior of the yurt is yet another work of art. The steppe carpet is not a pile or a flat weave one; it is a multi-coloured felt one, which decorates both the floor and walls of the yurt.[12] During the long winter months, Kyrgyz women masterfully embroider and weave a wide range of utility items, all of which exhibit a great artistic skill.

The hearth, situated directly beneath the *tunduk*, is the veritable heart of the yurt. Here, over a yak dung-fueled fire, the nomads' simple yet nutritious soups and stews are cooked in a copper *kazan* (pot).[13]

12  Nomads have inhabited the steppes of Eurasia for millenia. Before more complex weaving techniques were discovered and employed, the nomads of the steppes used a very simple method to produce practical, yet durable and attractive, felt carpets. Felt is made from pressed sheep, goat or camel wool. Rolls of felt, once compressed, are spread out and richly decorated by experienced tribal artists who stitch colourful patterns onto them. The finished felt carpets are then placed on the floors and walls of the yurt. Today, Kyrgyz nomads use knotted pile and flat weave techniques to an astonishing effect in their weaving. But the simple thick felt matting is still a favourite for yurt door covers and flooring.

13  Customarily, this cooking pot is never empty, but always full of hot and welcoming food. After meals, nomads would offer the following prayer: "May this home never be without food."

In 1992, when Kyrgyzstan became an independent republic, the *tunduk* was chosen as the central symbol on the new national flag. The red background symbolises bravery and valour, the sun peace and wealth while the *tunduk* represents the family home.

14 This opposition is based partly on the assumption that the nomads could never be fully integrated into the modern Kyrgyz society because Afghan Kyrgyz nomads have weak language skills.

A large section of the yurt is screened off and used for storing large vats of fresh yak milk and sacks of *quroot*. Another area is retained as a depot for hunting gear like saddles, harnesses and bridles. Women and children occupy a special corner, where baby cradles and wooden chests full of valuable items are kept. It is from such boxes that gifts are chosen for every departing guest, so that no one may ever leave the yurt empty-handed.

As mentioned previously, a small group of Kyrgyz nomads refused to follow their khan into exile in Türkiye in the 1980s. Today, several serious problems confronting the Kyrgyz nomads of the Wakhan Corridor might possibly embolden them to actually leave the 'roof of the world'. Infant mortality is extremely high as there are no clinics or hospitals in the region. This is exacerbated by a demographic imbalance: the unsustainable decreasing female population, which accounts for only a third of the total nomadic population of the Corridor. Since 1999, the government of neighbouring Kyrgyzstan have shown some token interest in repatriating their ethnically and linguistically related Turkic brethren to the 'motherland'. However, it is doubtful if they are indeed willing to allocate resources for this repatriation. In addition, many citizens of Kyrgyzstan are opposed to the possible relocation of 1,500 Afghan Kyrgyz nomads to their country.[14]

For many Afghan Kyrgyz, their decades-long search of a safe haven was realised once they were flown to Türkiye in 1982. But what they had to abandon forever was the freedom of their nomadic past. Today, they are living as full citizens of the Turkish Republic in a government-built settlement called *Ulupamir Köyü*

('Big Pamir Village'), near Van, in the east of the country.[15] The permanent community has a school, health clinic, post office, police station and mosque. A second generation of Kyrgyz are now being fully integrated into Turkish society. Sometimes, for the older generation – the parents and grandparents – their nomadic past is but a fond memory. They no longer practise their pastoral lifestyle although some Kyrgyz youngsters often hire themselves out as seasonal shepherds to local Turks.

As of 2016, the Kyrgyz families in the high Pamir plateaus of Afghanistan continue to survive against ever more life-threatening odds. Will they remain there and adapt to the new reality that is affecting their long-term existence, or will they succumb and opt for asylum in Kyrgyzstan or Türkiye, thus forever renouncing their nomadic lifestyle?[16]

> The Kirghiz odyssey is indeed a sad commentary on the plight of millions of nomadic pastoralists who, for the sake of their cultural integrity, managed to adapt for centuries to extremely unfriendly natural environments, only to be destroyed by the revolutions of this century [twentieth] which, ironically, promised or promise to liberate humanity.[17]

---

15  Ulupamir Village now has a population of some 2,500 (2012), a sizeable demographic growth.

16  Kazemi, "On the Roof of the Afghanistan: the Last Kyrgyz in Afghanistan."

17  Shahrani, "Afghanistan's Kirghiz in Türkiye."

The Kyrgyz yurt is sustainable as only natural readily available materials are used in its construction: wood for the frame and felt (pressed wool) for coverings. No nails are used as all parts are perfectly crafted to fit together. These materials offer excellent insulation for making the yurt cool in summer and warm in winter. The warm hearth is central and always has a hot pot of food awaiting any guest.

# 11
# ALTAI KAZAKHS
## OF NORTHWESTERN MONGOLIA

Of all the five former Soviet Central Asian republics Kazakhstan is geographically the largest (2,724,900 km²).[1] The country's nomadic ancestry is reflected in its official name. One etymology speculates that "kazakh" derives from the ancient Turkic word, *qaqzaq* ("to wander") and the Persian suffix-*stan* ("country" or "land").

1   The remaining four are Kyrgyzstan, Tajikistan, Turkmenistan and Uzbekistan.

The Kazakhs are direct descendants of ancient Turkic and Mongol tribes who inhabited for millenia the steppes of Eurasia. Steppe peoples were nomadic herders, who were able to move uninterruptedly from the grasslands of Mongolia to the fertile pastures of the Ukraine. Open grasslands provided regenerating fodder for the vast nomadic armies of mounted warriors, whose supremacy across the steppes was dependent on permanent access to these. Urban cultures, necessitating the cultivation of edible plants, of course, posed an impediment to free-roaming nomads. Genghiz Khan, for example, always in search of optimum grazing for his horses, despised permanent settlements and farmland, which he believed denied him this.

Islam was first introduced to Central Asia by Arab Muslims in the seventh and eighth centuries, but the Islamisation of the region was intensified during the reigns of the Persian-speaking Samanids (819-999). The Kazakhs, like all steppe peoples, lived in portable dome-shaped yurts and completed seasonal migrations. By the late nineteenth century, the nomadic lifestyle of most Central Asian tribes was curtailed by waves of Russian farmers – encouraged by the Czar – to settle on pastureland, and to indoctrinate the indigenous peoples with Russian culture and customs.[2] Tragically, up to two-thirds of the Kazakh population was annihilated during the bloody Russian Civil War (1917-1922). Many successful Kazakhs fled with their livestock to western China,[3] and those who remained in Kazakhstan were forcibly settled on Soviet collective farms.

Frontiers demarcating the independent states of Central Asia were not clearly defined until the twentieth century. For example, for hundreds of years Kazakh nomads roamed freely across

2   "Kazakh culture in Bayan Olgii, Western Mongolia." *Mongolia-Travel-Advice.com*.

3   Today, there is a sizeable Kazakh minority in the Uyghur Autonomous Region of Xinjiang, in western China. Smaller minorities inhabit parts of Iran, Russia, Uzbekistan and Mongolia. Though, it is heartbreaking how history is repeating itself with the destruction of nomadic cultural heritage in Xinjiang that has amassed from the incomprehensible captivity of Uyghur Muslims by the Chinese authorities since 2017.

Altai Kazakhs of Northwestern Mongolia 135

The Kazakhs of Mongolia, sequestered in the mountainous northwest, are not only geographically remote, but, being Turkic, are also linguistically distinct from the rest of the Mongolian-speaking population. This relative isolation has, in fact, enabled the Kazakhs to safeguard many of their ancient time-honoured traditions and skills. For example, the Kazakhs train large birds of prey (seen here soaring above the encampment) and hunt with them. The art of eagle hunting is believed to have been practised by them for hundreds of years.

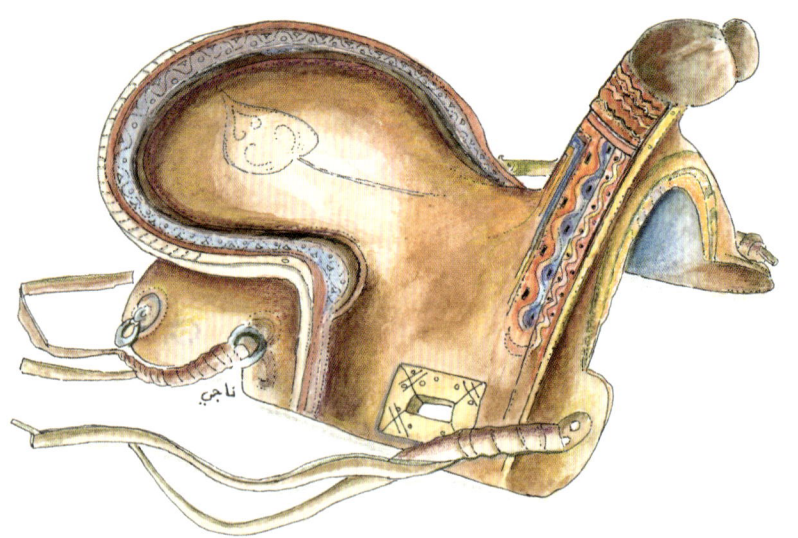

◀

Nomads over the world share a deep affection for the horse. The Kazakh nomads of Mongolia are skilled horsemen and hunters of prey. They make beautiful, exquisitely-tooled leather saddles.

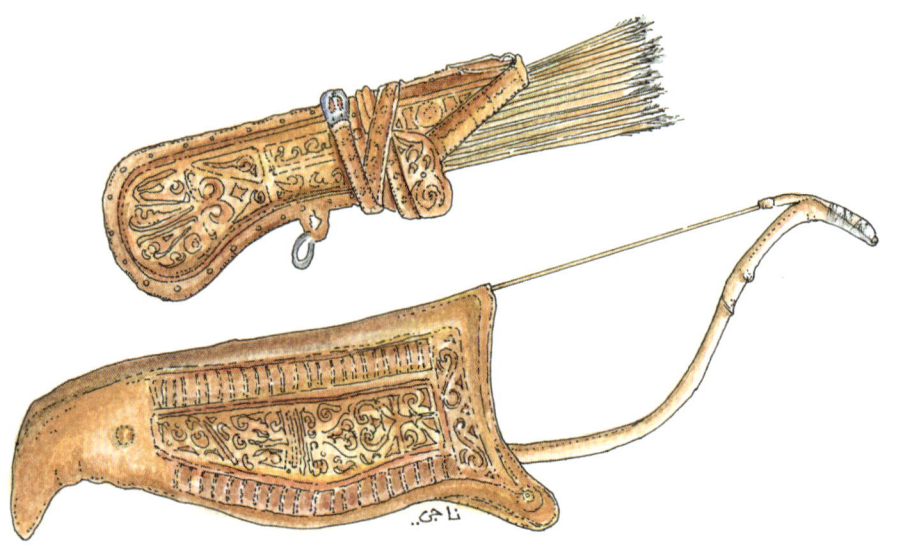

◀

The Kazak nomads of the Altai Mountains of western Mongolia are skilled hunters. They use well-trained golden eagles to catch prey; they also use bows and arrows. Durable bow cases and quivers are made of finely-tooled hide; their design has essentially remained the same since ancient times.

the borders of modern-day China, Kazakhstan and Mongolia. It was no surprise, that when the best grazing land and wells were confiscated by Russian farmers, many nomads chose to flee Kazakhstan after failed uprisings. Intent on preserving their nomadic lifestyle and religious beliefs, one group escaped east across the Altai Mountains to seek asylum in western Mongolia. Ironically, in 1924, Mongolia officially became the second Communist country in the world. Despite inhabiting the remote mountain valleys – very distant from the Mongolian capital, Ulaanbaatar – the Kazakh nomads still suffered from the purges and forced collectivisation of the new Communist regime, but to a lesser degree than their brethren in Kazakhstan.

Although the Kazakhs form the largest minority, numbering more than one hundred thousand in a country of almost three million inhabitants, they have not fully integrated into Mongolian society. They are determined to preserve their Islamic identity, culture and lifestyle that differ significantly from that of the majority of Mongolians. While slightly more than half of the population profess Buddhism, almost forty percent consider themselves non-religious. The Kazakhs of Mongolia, sequestered in the mountainous northwest, are not only geographically remote, but, being Turkic, are also linguistically distinct from the rest of the Mongolian-speaking population. This relative isolation has, in fact, enabled the Kazakhs to safeguard many of their ancient time-honoured traditions and skills.

Mongolian Kazakhs are traditionally nomadic pastoralists. Western Mongolia is comprised of a variety of eco-regions: mountains, forests, grassland steppe and deserts. Areas that receive insufficient rainfall for farming have historically been used by nomadic herdsmen. Under Communist rule (1924-1989), the pastoral economy of the country was collectivised, but today nomads are returning to their migratory ways, herding their livestock from one pasture to the next.

Kazakh nomads are not aimless wanderers; they move for a purpose. To successfully pursue their chosen lifestyle, they have invested everything they own and travel with all their possessions. The real wealth of a nomad lies in the animals he owns. It is only a very skilled leader who can lead his nomadic group to pastures with sufficient grazing and water supply.[4]

> Flocks and herds that might appear too small adequately to graze their pastures in summer might prove too large for the more meagre winter pastures. Yaks and camels, horses and mules, sheep and goats ... all required different grazing conditions and degrees of shelter from the winter blasts. Survival required a degree of mutual dependency; a single family would do well to own different varieties of stock as an insurance against disease and misadventure, but the family might need to farm out these varieties for grazing in different regions where the particular vegetation and cover provided the required fodder and protection. In reality little was haphazard about the nomadism of the steppes of Central Asia.[5]

Seasonal migration is the essence of the nomad's lifestyle. This "migration between two pastures is the most efficient use of marginal land yet devised, and settlement means that livestock can only be supported for part of the year..."[6] It is clearly necessary, for the survival of both man and beast,

---

4  This skill is acquired over time and is gained from much experience. The 'skilled' nomad, for example, can interpret signs in nature to determine which migratory routes to follow. An in-depth understanding of terrain and local weather patterns also ensures success.

5  Ure, *In Search of Nomads*, 154-155.

Long utilitarian *malbands* (wide tent straps) are woven by Kazak women on horizontal looms. They can be 10-15 metres in length and are wound around tent poles to reinforce the yurt.

that the summer grazing period be maximised as brutal winter temperatures in western Mongolia can fall to -40º C. The migratory patterns of families vary depending on the quality of the grass in a particular pasture. While some nomads might choose to migrate twice a year (once in the spring to reach the summer pasture and once again in the fall to reach the winter quarters), others might need to herd their animals to different pastures during each of the four seasons. As nomads will frequently return to the same grazing sites, the rights to these are often passed down within families.

Kazak nomads, like all Central Asian pastoralists, live in the ubiquitous circular yurt. This portable dwelling – sometimes transported on wheeled carts – can be quickly disassembled and loaded onto yaks or camels;[7] the felt coverings, wooden door, lattice walls, poles and wooden roof ring (*tunduk*) can be then easily reassembled. It is the ideal home for migratory nomads. All family members work in unison to ensure the continuance of the traditional way of life. Girls will busily churn mare's milk to make *kumis*, a staple of the Kazak diet. Women will spend hours drying balls of dried cheese (*quroot*). Daily meals are cooked over a dung-fueled fire in the central hearth. One traditional dish widely eaten is *bashparmak* ("five fingers"), a tasty meat and noodles stew.

Kazaks are extremely artistic and their tribal weavings that decorate the interior of the yurt reflect this. Frequently, beautifully-crafted circular felt carpets cover the entire floor. This carpeting is not only attractive, but also extremely functional. The thick felt, made of layers of pressed wool, is a natural insulator.[8] Long tent straps (*malbands*), wound around supportive tent poles, are woven on horizontal looms. Extremely durable rope is hand-made by braiding horse-hair.

For steppe nomads, the horse – believed to have been first

6  Tompson, *Carpets from the Tents, Cottages and Workshops of Asia*, 79.

7  Ibn Battuta (1304 – 1368/69), the mediaeval Moroccan Muslim traveller, passed through Central Asia and witnessed nomads on the move across the steppes. "…we saw a vast city on the move with its inhabitants, with mosques and bazaars in it, the smoke of the kitchen rising in the air (for they cook while on the march), and horse drawn wagons transporting the people. On reaching the camping place they took down the tents from the wagons and set them on the ground, for they are light to carry and so likewise they did with the mosque and shops." (Dunn, Ross E. *The Adventures of Ibn Battuta*. University of California Press, 2005)

8  Felt, compacted, not woven woollen fibres, is likely the oldest woollen fabric made by man; its origin is probably the heartland of the Turks: Central Asia and Mongolia. Nomads still make use of this very basic, yet durable material. The Kyrgyz insulate their yurts in winter with thick felt rugs and felt coverings weatherproof their exteriors.

In mediaeval times, the most common form of travel across the Eurasian steppes was the cart tent. The fourteenth century traveller, Ibn Battuta, describes the portable Central Asian yurts that were placed on four-wheeled carts: " ... There is placed upon the wagon a kind of cupola made of wooden laths tied together with thin strips of hide; this is light to carry, and covered with felt or blanket cloth, and in it there are grilled windows."

142  ENCOUNTERS WITH MUSLIM NOMADS

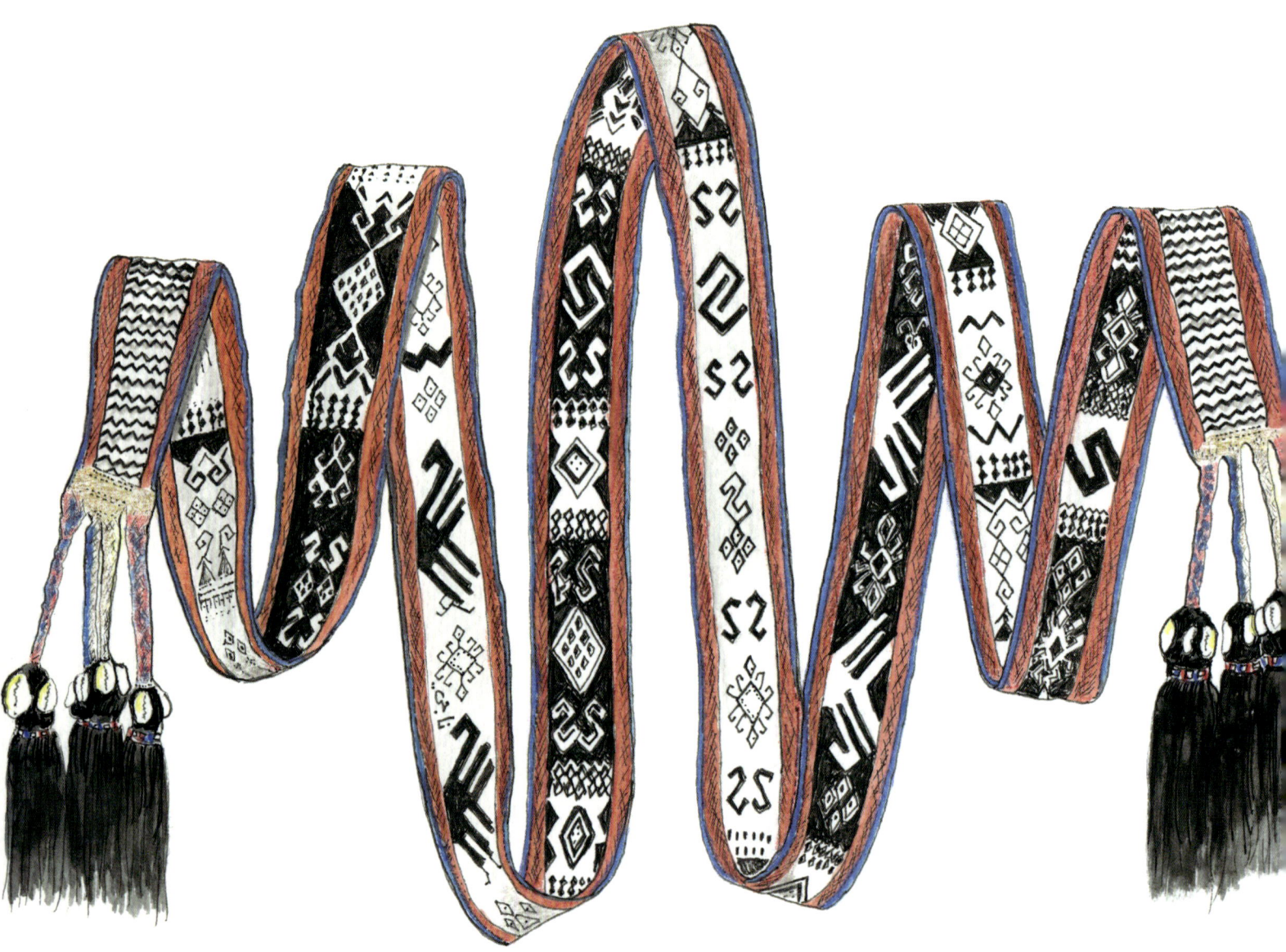

Narrower *malbands* are used to secure heavy loads on beasts of burden or even to strap in an infant atop a camel during a migration. For whatever utilitarian purpose they are used for, *malbands* are exquisite examples of tribal art.

domesticated on the steppes – was as important as the camel (the 'ship of the desert') was to the desert Bedouins.[9] It provided comfortable transportation, but was also a source of meat and milk. A Kazakh proverb states that "it is not the man who is reproved who dies, but the man who loves his horse."[10]

The Kazakh nomads of Mongolia are skilled horsemen and hunters of prey. Nomads over the world share a deep affection for the horse. It is well-known that the Arabian stallions, originally bred in the Arabian Peninsula, are some of the finest and most sought after of all breeds. The Kazakhs share another interest with Arab Bedouins: the ancient practice of training birds of prey and hunting with them. While the Bedouins of the Arabian Peninsula prefer falcons, the Kazakh nomads use a much larger bird, the golden eagle. The art of eagle hunting is believed to have been practised by them for hundreds of years.[11]

The training of the large golden eagle is an art passed down from father to son.[12] Female eagles are trained to travel on the thick padded arm rest of the nomad. When the hunter spots some prey (a marmot, fox or wolf), he releases the bird's headpiece, which covers the bird's eyes. Once the hunter is sure the eagle has spotted the prey, it is released. Fox fur and other animal pelts are used by the nomads to make warm winter hats and coats. Two animals are revered by the Kazakhs who say:

## "Fine horses and fierce eagles are the wings of the Kazakhs."

Not unlike the Kyrgyz nomads of Afghanistan's Wakhan Corridor, who were promised repatriation to motherland Kyrgyzstan, the Kazakhs of Mongolia were invited by the Soviet-

9   Horses are 'broken in' at an early age and are often ridden without saddles, at first. The mares, in particular, because they provide a daily source of fresh milk, soon become an integral part of the nomad's family.

10  Kazanov, *Nomads and the Outside World*, 47.

11  While the sport of hunting with eagles is perhaps a thousand years old, falconry has been practised in Central Asia for six millennia.

12  A female golden eagle can weigh up to 7 kilos and have a wind span of 2.2 metres.

backed Kazakh leader, Nursultan Nazarbayev, to return to their historic homeland. While the Kyrgyz government have not been willing to fund the mass transportation of hundreds of nomads, Nazarbayev's scheme has been in full operation for several decades. However, it has not been totally successful. Nazarbayev, with his country's enormous oil revenues, has built an ultra-modern society based on Kazakh nationalism. In his view, there is no place for the antiquated lifestyle represented by the Kazakh nomads of Mongolia.

A new repatriation programme, the "Blessed Migration", was inaugurated in 2009. However, Kazakh nomads who were initially lured to Nazarbayev's Kazakhstan by the substantial financial incentives are now having second thoughts, and some have chosen to return to live as before as part of a minority community in Mongolia. Many Kazakhs who have returned complain of blatant discrimination against them.[13]

> It has dawned on many that the siren song of nationalism means less in the long run than the company of Mongolians who, however different in language, ethnicity and religion … still find a place in their lives for yurts … for flocks of animals, for fiery little horses ridden bareback and all the other customs and consolations of life on the plains.[14]

13  Lillis,"Kazakhstan: Astana Lures Ethnic Kazakh Migrants with Financial Incentives."

14  Popham, "Wanders never cease: The nomadic life of Mongolia's Kazakh eagle hunters."

Altai Kazakhs of Northwestern Mongolia    145

The Kazak nomads are extremely artistic. Centuries old traditional patterns
are repeated on felt mats and on colourful tooled leather boots.

# 12

# FUTURE
## OF NOMADISM

Nomadism seems to have originated five thousand years ago in Asia with the advances in both animal husbandry, the development of practical portable housing, and later, the domestication of the horse. The Asian steppes were soon populated with tribes all sharing the same nomadic lifestyle.

Over the centuries, confederations of nomadic groups led by militant chieftains became formidable foes that created empires whose longevity depended on the strength of the social solidarity ('*asabiyyah*) that existed within the tribes. Islam, initially revealed to a nomadic people in Arabia, had spread within a century to the borders of Central Asia. Here, Turkic nomads embraced Islam and became fervent believers. They later founded the Ghaznavid and Seljuq Islamic dynasties, extending the frontiers of Islamic lands further east into India and west, into Byzantine-held Anatolia. To some extent, the present-day sedentary populations of most regions identified in this book have a nomadic ancestry. Despite all the attempts to sedentarise tribal peoples, nomadism continues to be the preferred lifestyle of select, relatively small Muslim populations inhabiting remote corners of the Islamic world.[1]

Despite their marginalisation in the twenty-first century, nomads continue to be an essential element in many urban economies, such as those of Iran and Afghanistan. Yet, researchers who recently investigated nomadic issues were polarised in their views about the true value of nomadism today. There were those who deemed pastoral nomads to be "culturally backwards, wasteful and economically destructive." Conversely, there were others who recognised the positive role of nomads in both "production and conservation."[2] Still, others viewed nomads as having an ancient anachronistic, yet fascinating lifestyle – one that is at the same time both primitive and embarrassing.[3] Some countries are, in fact, capitalising on their indigenous nomadic populations. For example, in Morocco, Iran and Jordan, tourist brochures now feature photos and information boxes highlighting the beauty and exotic nature of their indigenous

---

[1] Perhaps there is some truth in Charles Darwin's claim that the migratory urge is the strongest of animal instincts. (*Descent of Man*, chapter 4)

[2] Tapper, "Who are the Kuchi?", 106.

[3] Foschini, "The Social Wandering of Afghan Kuchis."

nomadic cultures. For the well-seasoned and inquisitive foreign tourist, package 'nomad' tours are available that guarantee an 'authentic experience' living with pastoralists in the mountains or deserts. And in post-revolutionary Iran, nomadism has been 'naturalised'; in an Iranian government publication posted in 1998, one article mentioned the start of the autumn migration of the Luri tribes while an adjacent one discussed the arrival in the country of grey cranes from Siberia.[4]

Pastoral nomadic cultures proverbially leave few traces of their existence, unlike sedentary societies that leave an abundance of artefacts for archaeologists to assay. Mobility for the nomad is crucial to his survival, so his worldly possessions are few; if needed, a nomad can quickly dismantle his tent or yurt and migrate to new quarters. The number of livestock determines his material wealth. However, nomads are custodians of millennia-old environmental knowledge. This is their true wealth that is in danger of being lost forever if and when a nomadic society is permanently settled, or if the indigenous language becomes extinct.

Traditional knowledge of the environment survives in the languages of nomadic peoples. For example, the Yup'ik of northeast Siberia and Alaska were semi-nomadic hunter-fisher-gatherers, who until the 1970s moved seasonally in search of sea mammals.[5] Global warming is melting the Arctic ice and making their ancestral hunting practices no longer possible. They have now become the first climate-change refugees. The Yup'ik possess a complex survival technology, as accurate and complex as any devised by modern science, that has enabled them to survive for over 6,000 years in one of the

4   Tapper, "Who are the Kuchi?, 109.

5   The Yup'ik are not Muslims, but their situation reflects the plight of many Muslim nomadic peoples today. Sedentarisation, forcible or willing, is a reality amongst most tribes today. Once nomads are settled, their languages are invariably discarded by the younger generation who prefer the language of the sedentary majority.

world's harshest environments. Their sophisticated knowledge of weather patterns of ice and snow, honed over generations of careful observation, make them amazingly accurate weather forecasters. This knowledge – scientific in its attention to detail – centres around winds, ocean currents and ice movement. By interpreting these key factors, the Yup'ik nomads are better able to evaluate the conditions for their important maritime hunting. This immense and invaluable repository of Arctic weather forecasting is encoded in detail in the Yup'ik languages. As these are now endangered, we face the possible loss of one of the most sensitive systems in existence to identify weather patterns and climate change, and also the loss of any clues that might help us in dealing with the causes and effects of global warming.[6]

The nomadic populations of western Mongolia, western China and Afghanistan have a bleak future. A century ago, the vast majority of Mongolia's tiny population were nomads; today, approximately thirty percent of its inhabitants are nomadic or semi-nomadic.[7] Although Mongolia is still one of the least densely populated nations on earth, accelerated modernisation schemes have had a deleterious effect on its nomadic population. For example, the exploitation of the country's mineral resources is threatening its pristine environment, especially the headwaters of its rivers. Overgrazing of pastures, climate change and extremely severe winters – resulting in the excessive loss of livestock – have forced many nomads to abandon their ancient lifestyle of the steppes for a more settled urban one. With the radical demise of this age-old culture, one hopes that history and tradition might still be valued in rapidly changing Mongolia.

6  Harrison, *The Last Speakers*, 72-77.

7  Mongolia's population is three million (2015), with more than sixty percent inhabiting the capital, Ulaanbaatar, and half a dozen smaller towns.

The minority Kazakh, Kyrgyz and Tajik nomadic populations of western China are faring no better than those of Mongolia. China, still the world's foremost polluter, is actively involved in monopolising the lucrative renewable energy market by producing wind turbines, for example, that now outsell those made in Denmark, one of the originators of the technology. According to Chinese propaganda, the size of nomads' livestock must be reduced or sold off entirely in order for the government to realise the sustainable development of the grassland. To this end, the Chinese have actually fenced off entire tracks of nomadic pasture to prevent grazing. Ecologists, however, believe this is a destructive practice as it does not prevent soil erosion. They maintain that grazing is a natural component of the ecosystem and actually contributes to its sustainability. Nomads are being encouraged to opt for subsidised housing in resettlement camps. Rather than forego their preferred lifestyle, many of the nomads of western China have sought asylum in neighbouring Kazakhstan.[8]

Despite the ongoing state of war, the Kuchis, along with other nomadic tribes of Afghanistan, are experiencing hardship due to government intransigence regarding the status of nomads in the twenty-first century. It appears that "... because few officials really believe that nomadism can find a place in modern Afghan society, the government has been unable to either rationalise nomadic pastoralism and help its partial survival under improved conditions or to effectively plan and assist the settling of those nomads who cannot cope with a migratory lifestyle."[9]

8  Sabrie, "The Last Nomads."

9  Foschini, "The Social Wandering of the Afghan Kuchis."

What *is* the future of nomadism in the Islamic world? The history of civilisations is replete with wars for prestige, aggrandisement and the appropriation of land. The known world that encompasses the six inhabited continents has been partitioned, frontiers drawn and the sedentary lifestyle encouraged. The wars that have resulted in one man's coveting of another's land have resulted in a catastrophic loss of life.[10] Clearly, a case can be made for the philosophy of nomadism that "recognises that land belongs to everyone and to no one."[11] Nomads, once able to roam freely throughout countries and uninterruptedly across manmade borders, are now feeling constrained.[12] It would be a great loss for humankind if all nomads chose or were forced to lead a sedentary existence. As the global population approaches seven billion, we are being forced to confront our reckless behaviour and admit that we are in large partly responsible for the dire state of the planet. We are desperately seeking solutions to our environmental problems and have universally agreed that these must be sustainable. Modern capitalist societies, encouraging unrestricted consumption, are complicit in an egregious waste of natural resources. And excessive consumerism, in turn, has led to societies being totally alienated from the real purpose of life. Sociologists, economists and others are suggesting a 'minimalist' lifestyle as a possible alternative.[13] The fact is that if we are to survive as a species, we must adopt a more ethical, environmentally friendly lifestyle – indeed a more minimalist one. As Western societies collectively realise this, they will indeed have embraced – to some degree – a 'nomadic mentality'. Furthermore, if we all do with less, and exhort conservation and not wastefulness, we will be closer to acknowledging many admonitions of our Creator and, at the same time, be able to appreciate and empathise with the nomads

10  In Islam, there are clear rulings regarding the wrongful seizure of property (*ghasb*), belonging to another by force. Such an act is considered *haraam* by consensus. "Normally, in Islamic law, any person who brings life to un-owned land by undertaking its cultivation or reclamation or otherwise putting it to beneficial use acquires it as private property. Only those actions which bring new life to the land confer ownership; mere exploitation does not constitute revival. *Ihya'* ['revival'] gives people a powerful incentive to invest in the sustainable use of the land and to provide for their welfare and the welfare of their families and descendants. However, lands in which development would be injurious to the general welfare are not acquired through *ihya'*." (Nagy, *Green Muslims*, 103.)

11  "History of Nomadism." *New Nomads*.

12  Nomadism recognises that land is a collective good that should be used for the benefit of all. ("History of Nomadism." *New Nomads*.)

13  According to the minimalist philosophy, happiness can be found outside of the hedonistic consumer culture, which only celebrates the amassing and importance of 'things'; each individual must determine what is essential and what is not in his life. An extension of this philosophy is the 'happiness movement'. For the past forty years, the tiny Himalayan kingdom of Bhutan has emphasised – as a measure of economic success – the evaluation of its citizens' 'happiness' (using the GNH [Gross National Happiness] and not their 'consumption' (using the standard GNP [Gross National Product]).

It would be a great loss for humankind if all nomads chose or were forced to lead a sedentary existence. We are desperately seeking solutions to our self-inflicted environmental problems and have universally agreed that these must be sustainable. Nomadism promotes self-sufficiency – a lifestyle that is appealing to more and more people who want to determine their own course. Muslim nomads have a very sustainable message for the twenty-first century urban dweller – one they have known for countless generations: what you *need* is far more important than what you may *want*. Perhaps the future of nomadism is a two-way symbiotic relationship, whereby nomadic and sedentary societies learn from each other. Here a Kazakh nomad has used sustainable solar panels to charge an automobile battery, electricity from which will power mainly light bulbs and portable radios.

in our midst. Nomadism promotes self-sufficiency – a lifestyle that is appealing to more and more people who want to determine their own course. Muslim nomads have a very sustainable message for the twenty-first century urban dweller – one they have known for countless generations: what you *need* is far more important than what you may *want*. This is what Muslims believe is precisely how Allah has enjoined them to live as His custodians of the Earth, sustainably sharing in its bounty.

> O children of Adam, take your adornment at every masjid, and eat and drink, but be not excessive. Indeed, He likes not those who commit excess. (7:31)

Muslim nomads travel light; what they collect and transport are the barest of necessities. Small is indeed beautiful. One only has to look inside a Bedouin or Bakhtiyari tent, or a Kazakh or Kyrgyz yurt to see the utilitarian beauty of their sustainable lifestyle, which has purposefully and proudly been on display in the illustrations and text of this book. Nomads, so often marginalised and seemingly forgotten by sedentary society, will continue to roam (albeit in much smaller numbers) the deserts, steppes and mountains of the Islamic world. They are indeed of "those who walk humbly upon the earth" (25:63).

# SELECT BIBLIOGRAPHY

The works listed here include the books and articles consulted in the writing of this book, and all sources referred to in the previous chapters and footnotes.

Abdullah, Aslam. "Quran's Message for Humanity." *IslamiCity*, 4 Dec. 2015. Accessed February 10, 2016. http://www.islamicity.org/6509/qurans-message-for-humanity/.

Akatay, Jane. "Yörük at Heart: Keeping the Nomad Heritage Alive." *Hürriyet Daily News*, 28 Feb. 2011, Accessed February 16, 2016. http://www.hurriyetdailynews.com/n.php?n=yoruk-2011-02-22.

Al-Ansari, Mohammed Jaber. *Encounter of History and Modernity: Khaldunism for a New Arab Culture*. Bloomington, IN: AuthorHouse, 2010.

Baharnaz, Mohammed Reza. *Nomads Migrating with Swallows*. Tehran: Gooya House of Culture & Art, 2007.

Baldauf, Scott. "Nomads in a Troubled Land." *Christian Science Monitor*, 9 Feb. 2004. Accessed April 25, 2016. http://www.csmonitor.com/2004/0209/p12s01-wosc.html.

Bashiri, Iraj. "The Kirghiz of Afghanistan." Anglefire.com. 2002. Accessed March 23, 2016. http://www.angelfire.com/rnb/bashiri/Kirghiz/Kirghiz.html.

Chatty, Dawn. *Nomadic Societies in the Middle East and North Africa Entering the 21st Century*. 2nd ed. Ledien: Brill, 2005.

"Syria's Bedouin Enter the Frey." *Foreign Affairs*, 13 Nov. 2013. Accessed April 15, 2016. https://www.foreignaffairs.com/articles/middle-east/2013-11-13/syrias-bedouin-enter-fray.

Cooper, Merian C. *Grass*. New York: G.P.Putnam's Sons, 1925.

Cootner, Cathryn M. *Anatolian Kilims*. London: Philip Wilson Publishers, 1990.

Digard, J.P. "Baktiyari Tribe." *Encyclopaedia Iranica*. 15 Dec. 1988. Accessed April 15, 2016. http://www.iranicaonline.org/articles/baktiari-tribe.

Dupree, Louis. *Afghanistan*. Princeton, NJ: University Press, 1974.

"Durable solutions for Kuchi IDPs in the South of Afghanistan: Options and Opportunities." *Asia Consultants International*, 15. For UNHCR Kandahar/Kabul, November 2006.

Engebrigtsen, Ada Ingrid, "Key Figure of mobility: the nomad." *Social Anthropology*, 25:1, 2017:42-54. Accessed October 3rd, 2023.

Foschini, Fabrizio. "The Social Wandering of the Afghan Kuchis." AAN Thematic Report, 04/2013. Accessed April 5, 2016. http://www.afghanistan-analysts.org/wp-content/uploads/2013/11/20131125_FFoschini-Kuchis.pdf.

Graham-Harrison, Emma. "Hidden Casualties of Afghan War." *The Guardian*. 9 Jan. 2013. Accessed April 20, 2016. http://www.theguardian.com/world/2013/jan/08/afghanistan-war-nomadic-farmers.

Grant, Richard. "Morocco: Walking with the Berber Nomads." *The Telegraph*, 9 Feb. 2013. Accessed March 5, 2016. http://www.telegraph.co.uk/travel/destinations/africa/morocco/articles/Morocco-walking-with-Berber-nomads/.

Harrison, David K. *The Last Speakers The Quest to Save the World's Most Endangered Languages*. Washington, D.C.: National Geographic, 2010.

Headley, Robert L. "People of the Camel." *Saudi Aramco World*. September/October 1964:2-4. Accessed January 10, 2016. http://archive.aramcoworld.com/issue/196405/people.of.the.camel.htm.

"History of Nomadism." *New Nomads*, 2015. Accessed March 13, 2016 http://www.new-nomads.net/history-of-nomadism.

Hull, Alastair and José Luczyc-Wynowska. *Kilim the Complete Guide*. London: Thames & Hudson Ltd., 1993.

Ibn Khaldun. *The Muqaddimah*. (trans. Franz Rosenthal). Vols. 1-3. London: Routledge and Kegan Paul, 1967.

Jereb, James F. *Arts and Crafts of Morocco*. London: Thames & Hudson Ltd., 1995. "Kazakh culture in Bayan Olgii, Western Mongolia." *Mongolia-Travel-Advice.com*. Accessed March 2, 2016. http://www.mongolia-travel-advice.com/kazakhs-in-western-mongolia.html.

"The Kazakh Yurta." *Kazenergy* No. 2 (26-27), 2009. Accessed March 14, 2016. http://www.kazenergy.com/ru/2-26-27-2009/3004-the-kazakh-yurta.html.

Kazemi, S. Reza. "On the Roof of the World: the Last Kyrgyz in Afghanistan." *Afghanistan Analysts Network*. 3 Nov. 2012. Accessed February 16, 2016. https://www.afghanistan-analysts.org/on-the-roof-of-the-world-the-last-kyrgyz-in-afghanistan/

Ker, Michelle and Jacob Locke. "Singing in the Wilderness: Kuchi Nomads in Modern Afghanistan." *Cornell International Affairs Review*. 3:2, 2010:1-2. Accessed February 13, 2016. http://www.studentpulse.com/articles/1260/singing-in-the-wilderness-kuchi-nomads-in-modern-afghanistan.

Khazanov, Anatoly M. *Nomads and the Outside World*. 2nd ed. Madison WI: The University of Wisconsin Press, 1994.

"Kuchi nomads seek a better deal." *Irin News*. 18 Feb. 2008. Accessed March 4, 2016. http://www.irinnews.org/report/76794/afghanistan-kuchi-nomads-seek-better-deal

Lillis, Joanna. "Kazakhstan: Astana Lures Ethnic Kazakh Migrants with Financial Incentives." *Eurasianet.org*. 26 Feb. 2009. Accessed February 8, 2016. http://www.eurasianet.org/departments/insightb/articles/eav022709.shtml

MacDonald, Brian W. *Tribal Rugs Treasures of the Black Tent*. Woodbridge, Suffolk: Antique Collectors' Club, 1997.

"The Mongolian Kazakhs." *The Mongolian Kazakh Disapora*. 2011. Accessed March 18, 2016. https://www.macalester.edu/academics/geography/mongolia/mongolian_kazakhs.html

Nagy, Luqman. *The Book of Islamic Dynasties*. London: Ta-Ha Publishers Ltd., 2008.

*Green Muslims*. Riyadh: Maktaba Dar-us-Salam, 2010.

*Ibn Khaldun*. Riyadh: Maktaba Dar-us-Salam, 2011.

"Nomadic Life & the Expansion of Islam." *Opposing Views*. Accessed January 26, 2016. http://people.opposingviews.com/nomadic-life-expansion-islam-9461.html

"The Nomadic Tribes of Arabia." *Boundless World History I: Ancient-1600*. Boundless, 21 Jul. 2015. Accessed February 3, 2016. https://www.boundless.com/world-history/textbooks/5652/middle-eastern-empires-the-late-classical-period-and-the-rise-ofislam-8/pre-islamic-arabia-42/the-nomadic-tribes-of-arabia-154-13223/.

Oberling, Pierre. "Qašqa'i Tribal Confederacy." Encyclopaedia Iranica. 20 July 2003. Accessed March 13, 2016.

Opie, James. *Tribal Rugs*. London: Laurence King Publishing, 1998. http://www.iranicaonline.org/articles/qasqai-tribal-confederacy-i.

Peters, John Durham. "Exile, nomadism, and diaspora: the stakes of mobility in the western canon", in J. Morra and M. Smith (eds.), *Spaces of visual culture*, London: Routledge, 2006

Petsopoulos, Yanni. *Kilims*. London: Thames & Hudson Ltd., 1979.

Pickering, Brooke, W. Russell Pickering, and Ralf S. Yohe. *Moroccan Carpets*. London: Hali Publications Limited, 1998.

Popham, Peter. "Wanders never cease: The nomadic life of Mongolia's Kazakh eagle-hunters." *The Independent*, 27 Jan.2013. Accessed April 3, 2016. http://www.independent.co.uk/arts-entertainment/art/features/wanders-never-cease-the-nomadic-life-of-mongolias-kazakheagle-hunters-8463784.html

Richardson, David and Sue. *Qaraqalpaqs of the Aral Delta*. Munich: Prestel Verlag, 2012.

Rickleton, Chris. "Kyrgyzstan: Kyrgyz Community in Afghanistan Looking for a Way Out." *Eurasianet.org*. 7 May 2012. Accessed February 3, 2016. http://www.eurasianet.org/node/65369

Roberts, Shawn and Jody Williams. *After the Guns Fall Silent: the Enduring Legacy of Landmines*. Cowley, Oxford: Oxfam Publishing, UK, 1995.

Rong, Jiang. *Wolf Totem*. New York: Penguin Books, 2008.

Russel, Gerard. "Why Nomads Win: What Ibn Khaldun Would Say About Afghanistan." *The Huffington Post*. 25 May, 2011. Accessed February 12, 2016. http://www.huffingtonpost.com/gerard-russell/why-nomads-win-what-ibn-k_b_447878.html.

Sabini, John. "The World of Islam." *Saudi Aramco World*, May/June 1976: 2-4.

Sabrie, Gilles. "The Last Nomads." gsabrie.com, Accessed March 16, 2016. http://www.gsabrie.com/lastnomads/

Shahrani, Nazif M. "Afghanistan's Kirghiz in Türkiye." *Cultural Survival Quarterly (CSQ)*, 8.1 (Spring 1984). Accessed March 14, 2016. https://www.culturalsurvival.org/ourpublications/csq/article/afghanistans-kirghiz-Türkiye+&cd=1&hl=en&ct=clnk&gl=t

*The Kirghiz and Wakhi of Afghanistan*. Seattle: University of Washington Press, 2002.

Shahshahani, Soheila. "Tribal schools of Iran: sedentarization through education." *Nomadic Peoples*, 36/37 1995: 145-156. Accessed March 12, 2016. http://cnp.nonuniv.ox.ac.uk/pdf/NP_journal_back_issues/Tribal_schools_of_Iran_S_Shahshahani.pdf.

Shor, Jean and Franc. "We took the Highroad in Afghanistan." *National Geographic*, November 1950. Accessed February 27, 2016. http://ngm.nationalgeographic.com/1950/11/wakhan-corridor/shor-te

Shuptar, Vitaliy. "The Kazakh Yurta." *Discovery-Kazakhstan.com*. 2008. Accessed March 12, 2016. http://www.discovery-kazakhstan.com/archive/2008/10_7.php

Southam, Hazel. "Morocco's Last Berbers on their 4,000-year-old Annual Migration: a Tradition Now under Threat." *The Independent*, 20 Aug. 2012. Accessed April 4, 2016. http://www.independent.co.uk/news/world/africa/moroccos-last-berbers-on-their-4000-year-old-annual-migration-a-tradition-that-is-now-under-threat-8063327.html.

Stark, Freya. *The Minaret of Djam*. London: Tauris Parke Paperbacks, 2010.

Steves, Rick. "Bedouins in Palestine – Nomads with a Permanent Address." *The Huffington Post*, 25 Jan. 2014. Accessed January 17, 2016. http://www.huffingtonpost.com/rick-steves/bedouins-in-palestine--no_b_4303666.html.

Stone, Caroline. "Ibn Khaldun and the Rise and Fall of Empires." *Saudi Aramco World*, Sept./Oct. 2006: 28-39.

Tanavoli, Parviz. *Persian Flatweaves*. Woodbridge, Suffolk: Antique Collectors' Club, 1998.

Tapper, Richard. "Who are the Kuchi?" Nomad self-identies in Afghanistan." *Journal of the Royal Anthropological Institute*. 14:1, March 2008: 96-116. Accessed March 18, 2016. http://onlinelibrary.wiley.com/doi/10.1111/j.1467-9655.2007.00480.x/abstract

Thompson, Jon. *Carpets from the Tents, Cottages and Workshops of Asia*. London: Laurence King Publishing, 1993.

Topham, John. *Traditional Crafts of Saudi Arabia*. London: Stacey International, 1981.

Towfiq, F. "'Ašayer (tribes)." *Encyclopaedia Iranica*. 16 August 2011. Accessed March 10, 2016. http://www.iranicaonline.org/articles/asayer-tribes.

Ure, John. *In Search of Nomads*. New York: Carrol & Graf Publishers, 2003.

Uslu, Fatih. *Bozahmwetli Yörük Asireti*. Konya: Çizgi Kitabevi, 2015.

"We have sent it down as an Arabic Qur'an so ..." *QuranicPath*. Accessed March 5, 2016. www.quranicpath.com/finerpoints/arabic_quran.html

Wertime, John T. *Sumak Bags of Northwest Persia & Transcaucasia*. London: Laurence King Publishing, 1998.

"World Directory of Minorities and Indigenous Peoples – Afghanistan: Kuchis". *Refworld* 2008. Accessed March 16, 2016. http://www.refworld.org/docid/49749d698.html